VOLUME TWO HUNDRED AND THREE

ADVANCES IN
IMAGING AND ELECTRON PHYSICS

VOLUME TWO HUNDRED AND THREE

ADVANCES IN IMAGING AND ELECTRON PHYSICS

Edited by

PETER W. HAWKES
CEMES-CNRS
Toulouse, France

ELSEVIER

ACADEMIC PRESS
An imprint of Elsevier

Cover photo credit:
The cover picture is taken from Fig. 1 of the chapter by J. Angulo (p. 12).

Academic Press is an imprint of Elsevier
125 London Wall, London EC2Y 5AS, United Kingdom
525 B Street, Suite 1800, San Diego, CA 92101-4495, United States
50 Hampshire Street, 5th Floor, Cambridge, MA 02139, United States
The Boulevard, Langford Lane, Kidlington, Oxford OX5 1GB, United Kingdom

Notices

Knowledge and best practice in this field are constantly changing. As new research and experience broaden our understanding, changes in research methods, professional practices, or medical treatment may become necessary.

Practitioners and researchers must always rely on their own experience and knowledge in evaluating and using any information, methods, compounds, or experiments described herein. In using such information or methods they should be mindful of their own safety and the safety of others, including parties for whom they have a professional responsibility.

To the fullest extent of the law, neither the Publisher nor the authors, contributors, or editors, assume any liability for any injury and/or damage to persons or property as a matter of products liability, negligence or otherwise, or from any use or operation of any methods, products, instructions, or ideas contained in the material herein.

ISBN: 978-0-12-812087-3
ISSN: 1076-5670

Publisher: Zoe Kruze
Acquisition Editor: Jason Mitchell
Editorial Project Manager: Shellie Bryant
Production Project Manager: Divya Krishna Kumar
Designer: Mark Rogers

Typeset by VTeX

CONTENTS

CONTRIBUTORS

Jesús Angulo
CMM – Centre de Morphologie Mathématique, MINES ParisTech, PSL Research University, Paris, France

A.B. Bok
Formerly Philips, Eindhoven and TNO-TH, Delft, Netherlands

H. de Lang
Formerly Philips, Eindhoven and TNO-TH, Delft, Netherlands

Clifford M. Krowne
Naval Research Laboratory, Washington, DC, United States

J.B. le Poole, deceased (1917–1993)

J. Roos
Formerly Philips, Eindhoven and TNO-TH, Delft, Netherlands

PREFACE

Two of the three chapters in this volume are unusual, in that they complement chapters in volume 202. J. Angulo explains another important aspect of mathematical morphology, the notion of convolution in (max, min) algebra. We are all familiar with convolution and its Fourier transform. This concept extends to mathematical morphology and has been explored in the past in connection with image algebra in particular. This new study sheds entirely new light on such convolutions.

The fact that structural modifications affect the critical temperature of superconductors is well known but the present interest in phenomena on the nanoscale require the question to be re-examined. C. Krowne, explores such modifications in detail and considers their implications for future developments.

Finally, we continue our program of reprinting articles from *Advances in Optical and Electron Microscopy*. The article by A.B. Bok, J. Le Poole, J. Roos, and H. de Lang remains the fullest account of early mirror electron microscopy, a subject that is currently in rapid development, and I am confident that many readers will be glad to have easy access to it.

As always, I am very grateful to the authors for their ability to make difficult subjects widely accessible.

Peter W. Hawkes

FUTURE CONTRIBUTIONS

S. Ando
Gradient operators and edge and corner detection

D. Batchelor
Soft x-ray microscopy

E. Bayro Corrochano
Quaternion wavelet transforms

C. Beeli
Structure and microscopy of quasicrystals

C. Bobisch, R. Möller
Ballistic electron microscopy

F. Bociort
Saddle-point methods in lens design

K. Bredies
Diffusion tensor imaging

A. Broers
A retrospective

A. Cornejo Rodriguez, F. Granados Agustin
Ronchigram quantification

J. Elorza
Fuzzy operators

R.G. Forbes
Liquid metal ion sources

P.L. Gai, E.D. Boyes
Aberration-corrected environmental microscopy

S. Golodetz
Watersheds and waterfalls

R. Herring, B. McMorran
Electron vortex beams

F. Houdellier, A. Arbouet
Ultrafast electron microscopy

M.S. Isaacson
Early STEM development

K. Ishizuka
Contrast transfer and crystal images

K. Jensen, D. Shiffler, J. Luginsland
Physics of field emission cold cathodes

U. Kaiser
The sub-Ångström low-voltage electron microscope project (SALVE)

K. Kimoto
Monochromators for the electron microscope

O.L. Krivanek
Aberration-corrected STEM

M. Kroupa
The Timepix detector and its applications

B. Lencová
Modern developments in electron optical calculations

H. Lichte
Developments in electron holography

M. Matsuya
Calculation of aberration coefficients using Lie algebra

J.A. Monsoriu
Fractal zone plates

L. Muray
Miniature electron optics and applications

M.A. O'Keefe
Electron image simulation

V. Ortalan
Ultrafast electron microscopy

D. Paganin, T. Gureyev, K. Pavlov
Intensity-linear methods in inverse imaging

N. Papamarkos, A. Kesidis
The inverse Hough transform

H. Qin
Swarm optimization and lens design

Q. Ramasse, R. Brydson
The SuperSTEM laboratory

B. Rieger, A.J. Koster
Image formation in cryo-electron microscopy

P. Rocca, M. Donelli
Imaging of dielectric objects

J. Rodenburg
Lensless imaging

J. Rouse, H.-n. Liu, E. Munro
The role of differential algebra in electron optics

J. Sánchez
Fisher vector encoding for the classification of natural images

P. Santi
Light sheet fluorescence microscopy

R. Shimizu, T. Ikuta, Y. Takai
Defocus image modulation processing in real time

T. Soma
Focus-deflection systems and their applications

J. Valdés
Recent developments concerning the Système International (SI)

J. van de Gronde, J.B.T.M. Roerdink
Modern non-scalar morphology

CHAPTER ONE

Convolution in (max, min)-Algebra and Its Role in Mathematical Morphology

Jesús Angulo
CMM – Centre de Morphologie Mathématique,
MINES ParisTech, PSL Research University, Paris, France
e-mail address: jesus.angulo@mines-paristech.fr

Contents

Advances in Imaging and Electron Physics, Volume 203
ISSN 1076-5670
http://dx.doi.org/10.1016/bs.aiep.2017.07.003

1. INTRODUCTION

Pioneered for Boolean random sets (Matheron, 1975), extended later to gray-level images (Serra, 1982) and more generally formulated in the framework of complete lattices (Serra, 1988; Heijmans, 1994), mathematical morphology is a nonlinear image processing methodology useful for efficiently solving many image analysis tasks (Soille, 1999).

Let E be the Euclidean $\mathbb{R}^n$ or discrete space $\mathbb{Z}^n$ (support space) and let $\mathcal{T}$ be a set of gray levels (space of values). For theoretical reasons it is assumed that $\mathcal{T} = \overline{\mathbb{R}} = \mathbb{R} \cup \{-\infty, +\infty\}$, but one often has $\mathcal{T} = [0, M]$. A gray-level image is represented by a function $f: E \to \mathcal{T}$, also noted as $f \in \mathcal{F}(E, \overline{\mathbb{R}})$, such that f maps each pixel $x \in E$ into a gray-level value in $\mathcal{T}$. Given a gray-level image, the two basic morphological mappings $\mathcal{F}(E, \mathcal{T}) \to \mathcal{F}(E, \mathcal{T})$ are the dilation and the erosion given respectively by

$$\begin{cases} (f \oplus b)(x) = \sup_{y \in E} \{f(y) + b(x - y)\}, \\ (f \ominus b)(x) = \inf_{y \in E} \{f(y) - b(y - x)\}, \end{cases} \tag{1}$$

where $b \in \mathcal{F}(E, \mathcal{T})$ is the structuring function which determines the effect of the operator. The other morphological operators, such as the opening and the closing, are obtained by composition of dilation/erosion (Serra, 1982; Heijmans, 1994). The Euclidean framework has been recently generalized to images supported on Riemannian manifolds (Angulo & Velasco-Forero, 2014).

If we compare the mathematical structure of the operators (1) to the classical convolution of a function f by a kernel k, let say the convolution in the $(+, \times)$-algebra; i.e.,

$$(f * k)(x) = \int_E f(y) k(x - y) dy,$$

then, we can establish in parallelism of the involved mathematical operations. Thus, operators (1) are interpreted in nonlinear mathematics as the convolution in $(\max, +)$-algebra (and in its dual algebra) (Gondran & Minoux, 2008). More precisely, the study of the operator equivalent to the erosion (1) can be traced back to the notion of inf-convolution (or infimal convolution) introduced by Moreau in the 70's of last century (Moreau, 1970), as the fundamental tool in convex analysis. Convolution in $(\max, +)$-algebra has been also widely studied in the framework of idempotent analysis developed by Maslov and co-workers (Kolokoltsov

& Maslov, 1997). This inherent connection of functional operators (1) with the supremal and infimal convolution of nonlinear mathematics and convex analysis has been extremely fruitful to the state-of-the-art on mathematical morphology. Morphological PDEs (Alvarez, Guichard, Lions, & Morel, 1993; Arehart, Vincent, & Kimia, 1993; Brockett & Maragos, 1994; Maragos, 1996), the slope transform (Dorst & van den Boomgaard, 1994; Maragos, 1995), or more recently to the notion of stochastic morphology based on Maslov random walks (Angulo & Velasco-Forero, 2013), are particularizations of results from nonlinear mathematics.

Nevertheless, the functional operators (1) do not extend all the fundamental properties of the dilation and erosion for sets, as formulated in Matheron's theory. Perhaps the most disturbing for us are, on the one hand, the lack of commutation with level set processing for non-flat structuring functions; on the other hand, the limitation of Matheron's axiomatic of granulometry to constant (i.e., flat) functions on a convex domain (Kraus, Heijmans, & Dougherty, 1993). In addition, there are some unconventional morphological frameworks, such as the fuzzy morphology (Nachtegael & Kerre, 2001; Deng & Heijmans, 2002; Maragos, 2005; Bloch, 2009) or the viscous morphology (Vachier & Meyer, 2005, 2007; Maragos & Vachier, 2008) which do not fit in the classical $(\max, +)$-algebra framework.

Actually, the $(\max, +)$ is not the unique possible alternative to see morphological operators as convolutions. The idea in this paper is to consider the operation of convolution of two functions in the $(\max, \min)$-algebra. This is in fact our main motivation: to formally introduce the notion of $(\max, \min)$-mathematical morphology. As we show in the paper, this framework is not totally new in morphology since some fuzzy morphological operators are exactly the same convolutions that we introduce. But some of the key properties are ignored by in the fuzzy context, and the most important, they are not limited to fuzzy sets. By the way, even if much less considered than the supremal and infimal convolutions, convolutions in $(\max, \min)$-algebra have been the object of various studies in different branches of nonlinear applied mathematics, from quasi-convex analysis (Volle, 1994, 1998; Seeger & Volle, 1995; Gondran, 1996; Luc & Volle, 1997; Penot & Zălinescu, 2001) to viscosity solutions of Hamilton–Jacobi equations (Barron, Jensen, & Liu, 1996, 1997; Alvarez, Barron, & Ishii, 1999; Van & Son, 2006). Interested reader is also referred to the book by Gondran and Minoux (2008) for a systematic comparative study of matrix algebra and calculus in the three algebras $(+, \times)$, $(\max, +)$

and (max, min), and to the book by Clarke and Stern (1999) for a nonlinear partial differential equations viewpoint. Results from this literature are extremely useful for us. Indeed, convolution in (max, min)-algebra is known in convex analysis literature as "sublevel convolution" or "level sets sum convolution". The reason for this name is obvious: the fundamental property of convolution in (max, min)-algebra is the commutation with level set processing. In fact, this principle is the inspiration and second motivation of this work: to show how (max, min) mathematical morphology can cover some of the unconventional operators which are formalized by level set processing. Let us precise that we do not exploit the strict axiomatic notion of "algebra", but the term is kept to emphasize the comparison with the alternative "algebras". Readers interested on more "algebra aspects" are refereed to related material in Ritter and Wilson (2011) and Maragos (2013).

The present work is exclusively a theoretical study and thus the practical interest of the operators is not illustrated here.

Paper organization. The rest of the paper is organized as follows.

- Section 2 gives an overview to basic notions on classical mathematical morphology and fixes the notation.
- A formal definition of morphological operators in (max, min)-algebra is introduced in Section 3. Relevant properties from an algebraic viewpoint are stated and proved.
- Sections 4 and 5 provide an overview of nonlinear analysis in (max, min) mathematics and can be straightforward related to our (max, min)-convolutions. On the one hand, the theory of viscosity solutions of the Hamilton–Jacobi equation with Hamiltonians containing u and Du is summarized, in particular, the corresponding Hopf–Lax–Oleinik formulas are given in Section 4. Section 5 discusses the results on quasi-concavity preservation, Lipschitz approximation and conjugate/transform related to (max, min)-convolutions.
- Section 6 reviews the links between (max, min)-convolutions and some previous approaches of unconventional morphology, in particular fuzzy morphology and viscous morphology. In addition, the interest of (max, min)-convolutions in Boolean random function characterization is considered. Links of (max, min) framework to geodesic dilation and erosion are also provided.
- Section 7 of conclusion and perspectives closes the paper.

2. BASIC NOTIONS AND NOTATIONS

For more details on this background material, see classical references on set morphology (Matheron, 1975), on flat morphology for functions (Serra, 1982, Chapter XII; Maragos & Schafer, 1987) or on complete lattice formulation of morphological operators (Serra, 1988; Heijmans, 1994).

Minkowski addition and subtraction. Given a set $X \subseteq E$, the complement of X is $X^c = E \setminus X$, and the transpose of X (or symmetrical set with respect to the origin O) is $\check{X} = \{-x : x \in X\}$. For every $p \in E$, the translate of X by p is $X_p = \{x + p : x \in X\}$. For any pair of sets X and Y, their Minkowski addition $\oplus$ and Minkowski subtraction $\ominus$ are defined as follows:

$$X \oplus Y = \bigcup_{y \in Y} X_y = \{x + y : x \in X, y \in Y\} = \left\{p \in E : X \cap \check{Y}_p \neq \emptyset\right\}, \quad (2)$$

$$X \ominus Y = \bigcap_{y \in Y} X_{-y} = \{p \in E : Y_p \subset X\} = \left\{x : \forall p \in \check{Y}, x \in X_p\right\}. \quad (3)$$

We remind that the (binary) dilation and erosion of set X by structuring element B, $B \subseteq E$, are just defined respectively as

$$\delta_B(X) = X \oplus B \quad \text{and} \quad \varepsilon_B(X) = X \ominus B.$$

Let us precise that we assume here and for the rest of the paper that B is a compact set. It should be noted that dilation and erosion are dual by complementation:

$$X \oplus B = \left(X^c \ominus \check{B}\right)^c \quad \text{and} \quad X \ominus B = \left(X^c \oplus \check{B}\right)^c.$$

It is sometimes misunderstood that duality by complement is the right way to construct erosion from dilation and vice versa. There is however another duality in mathematical morphology which is just indicated for such purpose: the duality by adjunction, which in the case of binary operators involves

$$X \oplus B \subseteq Y \quad \Leftrightarrow \quad X \subseteq Y \ominus B.$$

Composition of dilation and erosion leads to two extremely powerful morphological operators. Namely, the opening and the closing of set X by

B, which are defined respectively as

$$X \circ B = (X \ominus B) \oplus B, \quad (4)$$
$$X \bullet B = (X \oplus B) \ominus B. \quad (5)$$

Opening is anti-extensive and closing is extensive. But the most important property of the opening and closing operators is their idempotency; that is, their stability to iteration.

Erosion, dilation and adjunction in complete lattices. Morphological operators are formulated in a more abstract framework based on complete lattice theory. We briefly recall this framework.

Let us consider a nonempty set $\mathcal{L}$ endowed with a partial ordering $\leq$. We say that $(\mathcal{L}, \leq)$ is a complete lattice if every subset $X \subset \mathcal{L}$ has an infimum $\bigwedge X$ and a supremum $\bigvee X$ in $\mathcal{L}$. The least and the greatest elements of a complete lattice are denoted respectively by $\bot$ and $\top$.

The operators $\varepsilon : \mathcal{L} \to \mathcal{L}$ and $\delta : \mathcal{L} \to \mathcal{L}$ are called an erosion and a dilation if they commute respectively with the infimum and the supremum; i.e., $\varepsilon\left(\bigwedge_{i\in I} X_i\right) = \bigwedge_{i\in I} \varepsilon(X_i)$ and $\delta\left(\bigvee_{i\in} X_i\right) = \bigvee_{i\in} \delta(X_i)$, for every collection $\{X_i : i \in\} \subset \mathcal{L}$. Erosion and dilation are increasing operators, i.e., $\forall X, Y \in \mathcal{L}$, if $X \leq Y$ then $\varepsilon(X) \leq \varepsilon(Y)$ and $\delta(X) \leq \delta(Y)$. Erosion and dilation are related by the notion of adjunction:

$$\delta(X) \leq Y \Leftrightarrow X \leq \varepsilon(Y); \;\; \forall X, Y \in \mathcal{L}. \quad (6)$$

Adjunction law (6) is of fundamental importance in mathematical morphology since it allows to define a unique dilation δ associated to a given erosion ε, i.e.,

$$\delta(X) = \bigwedge\{Y \in \mathcal{L} : X \leq \varepsilon(Y)\}; \;\; X \in \mathcal{L}. \quad (7)$$

Similarly one can define a unique erosion from a given dilation:

$$\varepsilon(Y) = \bigvee\{X \in \mathcal{L} : \delta(X) \leq Y\}; \;\; Y \in \mathcal{L}. \quad (8)$$

Given an adjunction (ε, δ), their composition product operators, $\gamma(X) = \delta(\varepsilon(X))$ and $\varphi(Y) = \varepsilon(\delta(Y))$ are respectively an opening and a closing, which are the basic *morphological filters* having the following properties: idempotency $\gamma\gamma(X) = \gamma(X)$, anti-extensivity $\gamma(X) \leq X$ and extensivity $X \leq \varphi(X)$, and increaseness. Another relevant result is the fact that, given an erosion ε, the opening and closing by adjunction are exclusively defined in terms of this erosion as

$$\begin{aligned} \gamma(X) &= \bigwedge\{Y \in \mathcal{L} : \varepsilon(X) \leq \varepsilon(Y)\}, \quad \forall X \in \mathcal{L}, \\ \varphi(X) &= \bigwedge\{\varepsilon(Y) : Y \in \mathcal{L},\ X \leq \varepsilon(Y)\}, \quad \forall X \in \mathcal{L}. \end{aligned}$$

Dilation and erosion for functions and links with level set processing. As mentioned above, the dilation and erosion of a function $f \in \mathcal{F}(E, \overline{\mathbb{R}})$, by a structuring function $b \in \mathcal{F}(E, \overline{\mathbb{R}})$ are defined as

$$\delta_b(f)(x) = (f \oplus b)(x) \quad \text{and} \quad \varepsilon_b(f)(x) = (f \ominus b)(x),$$

where $(f \oplus b)$ and $(f \ominus b)$ were given in (1).

Dilation and erosion are dual by adjunction, i.e.,

$$f \oplus b \leq g \quad \Leftrightarrow \quad f \leq g \ominus b$$

as well as dual by complement, i.e.,

$$(f^c \oplus b)^c = f \ominus \check{b} \quad \text{and} \quad (f^c \ominus b)^c = f \oplus \check{b},$$

where the complement function ϕ^c of function ϕ is defined as the negation for real valued functions and the symmetric with respect to M for function valued in a nonnegative interval $[0, M]$, i.e., $\phi^c(x) = -\phi(x)$ if $\phi \in \mathcal{F}(E, \overline{\mathbb{R}})$ or $\phi^c(x) = M - \phi(x)$ if $\phi \in \mathcal{F}(E, [0, M])$. The transposed function $\check{\phi}$ is given by $\check{\phi}(x) = \phi(-x)$.

The structuring function is usually a parametric multiscale family $b_\lambda(x)$, where $\lambda > 0$ is the scale parameter such that $b_\lambda(x) = \lambda b(x/\lambda)$ and which satisfies the semi-group property $(b_\lambda \oplus b_\mu)(x) = b_{\lambda+\mu}(x)$.

The most commonly studied framework for dilation/erosion of functions, which additionally presents better properties of invariance, is based on flat structuring functions, therefore viewed as structuring elements. More precisely, given the structuring element $B \subseteq E$, its associated structuring function is

$$b(x) = \begin{cases} 0 & \text{if } x \in B \\ -\infty & \text{if } x \in B^c \end{cases}$$

Hence, the flat dilation $(f \oplus B)$ and flat erosion $(f \ominus B)$ can be computed respectively by the moving local maxima and minima filters; i.e., $(f \oplus B)(x) = \sup_{y \in \check{B}} f(x+y)$ and $(f \ominus B)(x) = \inf_{y \in B} f(x+y)$.

Given an upper semicontinuous (USC) function $f \in \mathcal{F}(E, \overline{\mathbb{R}})$, it can be defined by means of its upper level sets $X_h^+(f)$ as follows

$$f(x) = \sup\left\{h \in \overline{\mathbb{R}} : x \in X_h^+(f)\right\},$$

or if f is lower semicontinuous (LSC), using its lower level sets $X_h^-(f)$, as

$$f(x) = \inf\left\{h \in \overline{\mathbb{R}} : x \subset X_h^-(f)\right\}.$$

We note that a USC function can be also represented as

$$\begin{aligned} f(x) &= \inf\left\{h \in \overline{\mathbb{R}} : x \notin X_h^+(f)\right\} \\ &= \inf\left\{h \in \overline{\mathbb{R}} : x \in Y_h^-(f)\right\}, \end{aligned}$$

similarly an LSC function can be represented as

$$f(x) = \sup\left\{h \in \overline{\mathbb{R}} : x \in Y_h^+(f)\right\},$$

where

$$\begin{aligned} X_h^+(f) &= \left\{x \in E : f(x) \geq h\right\}, \quad \text{and} \quad Y_h^+(f) = \left\{x \in E : f(x) > h\right\}; \\ X_h^-(f) &= \left\{x \in E : f(x) \leq h\right\}, \quad \text{and} \quad Y_h^-(f) = \left\{x \in E : f(x) < h\right\}. \end{aligned}$$

Considering $E \subset \overline{\mathbb{R}}^n$, the upper level sets X_t^+ of any F are closed sets in $\overline{\mathbb{R}}^n$, decreasing, i.e., $h < k \Rightarrow X_k^+ \subseteq X_h^+$ and obey the monotonic continuity $X_k^+ = \cap_{h<k} X_h^+$. Lower level sets are closed sets, increasing $h < k \Rightarrow X_h^- \subseteq X_k^-$ and $X_k^- = \cup_{h<k} X_h^-$. Obviously, a continuous function f can be decomposed/reconstructed using either its (strict) upper level sets or its (strict) lower level sets. We note that, using duality by complement, one has

$$(X_h^+(f))^c = Y_h^-(f), \tag{9}$$

and thus the strict lower sets $Y_h^-(f)$ are open sets. We note also that a sampled function is trivially both USC and LSC because all its level sets are both closed and open sets.

As mentioned before, flat operators commute with level set processing. Being more precise, first notice that for any collection of continuous functions $\phi_i \in \mathcal{F}(E, \overline{\mathbb{R}})$, $i \in I$, we have (Serra, 1982, page 431):

$$X_h^+\left(\bigvee_i \phi_i\right) = \{x \in E : \phi_1(x) \geq h \text{ or } \phi_2(x) \geq h \text{ or } \cdots\} = \bigcup_i X_h^+(\phi_i), \quad (10)$$

$$X_h^+\left(\bigwedge_i \phi_i\right) = \{x \in E : \phi_1(x) \geq h \text{ and } \phi_2(x) \geq h \text{ and } \cdots\} = \bigcap_i X_h^+(\phi_i), \quad (11)$$

and by (9), their dual expressions by complement:

$$Y_h^-\left(\bigvee_i \phi_i\right) = \{x \in E : \phi_1(x) < h \text{ and } \phi_2(x) < h \text{ and } \cdots\} = \bigcap_i Y_h^-(\phi_i), \quad (12)$$

$$Y_h^-\left(\bigwedge_i \phi_i\right) = \{x \in E : \phi_1(x) < h \text{ or } \phi_2(x) < h \text{ or } \cdots\} = \bigcup_i Y_h^-(\phi_i). \quad (13)$$

The following formulas (Luc & Volle, 1997), relating lower level sets and strict lower level sets, are also useful in the sequel:

$$X_h^-(f) = \bigcap_{k>h} Y_k^-(f), \quad (14)$$

$$Y_h^-(f) = \bigcup_{j=1}^{\infty} X_{h-1/j}^-(f), \quad (15)$$

where the family $\left\{X_{h-(1/j)}^-(f)\right\}$, $j \geq 1$, is increasing.

Then, taking for ϕ_i in (10) and (11) all the translates of f as x runs over a set B or its transpose, together with definitions of $\oplus$ and $\ominus$, respectively (2) and (3), we can write:

$$\begin{aligned} X_h^+(f \oplus B) &= \{x \in E : \exists y \in B_x, f(y) \geq h\} = X_h^+(f) \oplus B, \\ X_h^+(f \ominus B) &= \left\{x \in E : \forall y \in \check{B}_x, f(y) \geq h\right\} = X_h^+(f) \ominus B. \end{aligned}$$

This property of commutativity of upper level sets involves that the flat dilation and erosion of a continuous function $f \in \mathcal{F}(E, \overline{\mathbb{R}})$ by structuring element $B \subset E$ are obtained as:

$$\delta_B(f)(x) = \sup\left\{h \in \overline{\mathbb{R}} : x \in \left(X_h^+(f) \oplus B\right)\right\}, \quad (16)$$

$$\varepsilon_B(f)(x) = \sup\left\{h \in \overline{\mathbb{R}} : x \in \left(X_h^+(f) \ominus B\right)\right\}. \quad (17)$$

For the opening and closing of a function f by a flat structuring element, denoted $\gamma_B(f)$ and $\varphi_B(f)$, and defined by

$$\begin{aligned} \gamma_B(f) &= f \circ B = (f \ominus B) \oplus B, \\ \varphi_B(f) &= f \bullet B = (f \oplus B) \ominus B, \end{aligned}$$

we also have a natural formulation using upper level sets:

$$\begin{aligned} X_h^+(f \circ B) &= X_h^+(f) \circ B, \\ X_h^+(f \bullet B) &= X_h^+(f) \bullet B. \end{aligned}$$

In the case of unflat dilation and erosion of function f by structuring function b, their formulation using sets is not very useful since it needs all the upper level sets:

$$X_k^+(f \oplus b) = \bigcup_h \left[X_h^+(f) \oplus X_{k-h}^+(b)\right], \quad (18)$$

$$X_k^+(f \ominus b) = \bigcap_h \left[X_h^+(f) \ominus X_{h-k}^+(b)\right]. \quad (19)$$

3. (max, min)-CONVOLUTIONS: DEFINITION AND PROPERTIES

In this section we define the alternative convolutions associated to a pair (*function* f, *structuring function* b) in the (max, min) mathematical framework. We also study their properties.

3.1 Definition

Definition 1. Given a structuring function $b \in \mathcal{F}(\mathbb{R}^n, \overline{\mathbb{R}})$, for any function $f \in \mathcal{F}(\mathbb{R}^n, \overline{\mathbb{R}})$ we define the supmin convolution $f \triangledown b$ and the infmax convolution $f \vartriangle b$ of f by b as

$$(f \triangledown b)(x) = \sup_{y \in \mathbb{R}^n} \left\{f(y) \wedge b(x - y)\right\}, \quad (20)$$

and

$$(f \vartriangle b)(x) = \inf_{y \in \mathbb{R}^n} \left\{f(y) \vee b^c(y - x)\right\}. \quad (21)$$

We also define the adjoint infmax $f \Delta^* b$ and the adjoint supmin $f \nabla^* b$ convolutions as

$$(f \Delta^* b)(x) = \inf_{y \in \mathbb{R}^n} \left\{ f(y) \wedge^* b(y-x) \right\}, \tag{22}$$

and

$$(f \nabla^* b)(x) = \sup_{y \in \mathbb{R}^n} \left\{ f(y) \vee^* b^c(x-y) \right\}, \tag{23}$$

where $\wedge^*$ is the adjoint operator to the minimum $\wedge$ and is given by

$$f(y) \wedge^* b(y-x) = \begin{cases} f(y) & \text{if } b(y-x) > f(y) \\ \top & \text{if } b(y-x) \leq f(y) \end{cases} \tag{24}$$

and $\vee^*$ the adjoint to $\vee$:

$$f(y) \vee^* b^c(x-y) = \begin{cases} f(y) & \text{if } b^c(x-y) < f(y) \\ \bot & \text{if } b^c(x-y) \geq f(y) \end{cases} \tag{25}$$

and where, if we define $\max g = \sup_{x \in \mathbb{R}^n} g(x)$ and $\min g = \inf_{x \in \mathbb{R}^n} g(x)$, the top and bottom elements for pair of functions f and b correspond to

$$\top = (\max f) \vee (\max b) \quad \text{and} \quad \bot = (\min f) \wedge (\min b^c).$$

Definitions remain valid if we replace $\mathbb{R}^n$ by a subset E or any subset of discrete space $\mathbb{Z}^n$. Similarly, the extended real line $\overline{\mathbb{R}}$ can be replaced by a bounded, eventually discrete, set of intensities $[0, M]$.

Fig. 1 illustrates the four (max, min)-convolutions for a given example of one dimensional functions defined in a bounded interval, i.e., $f, b \in \mathcal{F}(\mathbb{R}, [0, M])$.

The fact that four convolution operators are naturally defined in (max, min)-algebra is related to the previous discussion on dualities by complement and by adjunction and is easily justified by the following property.

3.2 Duality by Complement vs. Duality by Adjunction

It is obvious that the four operators are translation invariant. Furthermore, from a morphological viewpoint, their most salient properties are summarized in this proposition.

Proposition 2. *The supmin convolution ∇ and infmax convolution Δ are dual with respect to the complement. Similarly, the adjoint infmax convolution Δ^* and*

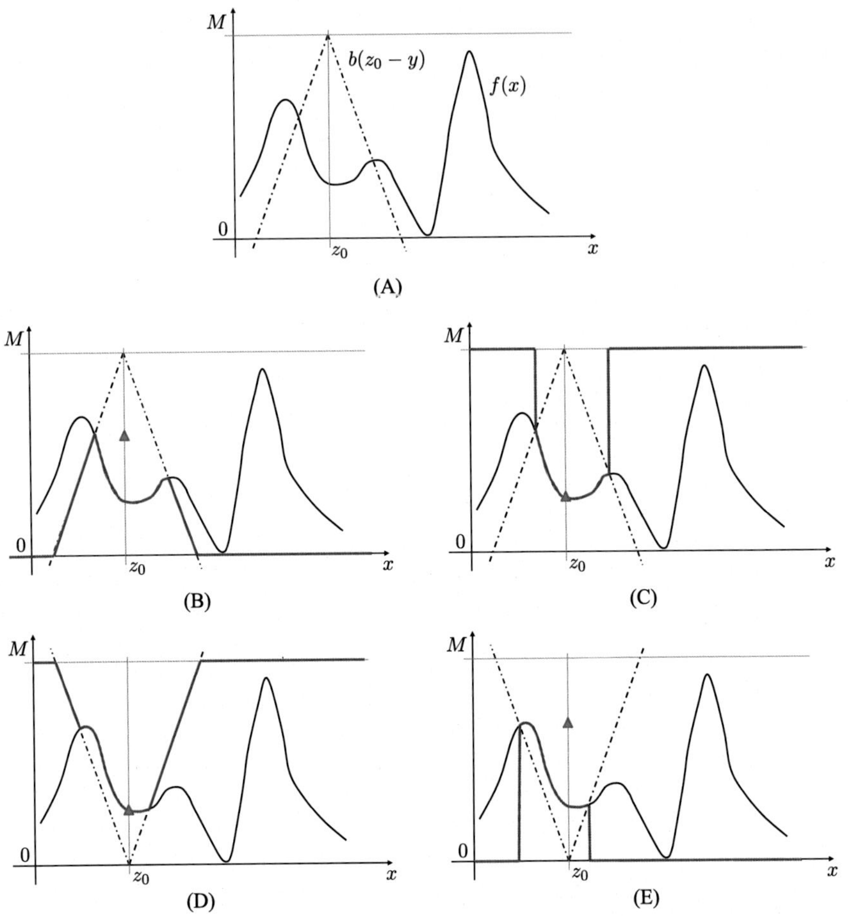

Figure 1 Illustration of four (max, min)-convolutions for a given example of one dimensional functions defined in a bounded interval, i.e., $f, b \in \mathcal{F}(\mathbb{R}, [0, M])$. (A) Original function $f(x)$ and translated structuring function b at point z_0; (B) in red, $f(y) \wedge b(z_0 - y)$ for all $y \in \mathbb{R}$, green triangle represents $(f \bigtriangledown b)(x)$ the value of the supmin convolution at z_0; (C) in red, $f(y) \wedge^* b(y - z_0)$, green triangle, adjoint infmax at z_0: $(f \bigtriangleup^* b)(z_0)$; (D) in red, $f(y) \vee b^c(y - z_0)$, green triangle, infmax at z_0: $(f \bigtriangleup b)(z_0)$; (E) in red, $f(y) \vee^* b^c(z_0 - y)$, green triangle, adjoint supmin at z_0: $(f \bigtriangledown^* b)(z_0)$.

the adjoint supmin $\bigtriangledown^*$ convolution are dual with respect to the complement, i.e., for $f, b \in \mathcal{F}(\mathbb{R}^n, \overline{\mathbb{R}})$ one has

$$f \bigtriangleup b = \left(f^c \bigtriangledown \check{b}\right)^c \quad \text{and} \quad f \bigtriangledown b = \left(f^c \bigtriangleup \check{b}\right)^c, \tag{26}$$

$$f \,\Delta^*\, b = \left(f^c \,\nabla^*\, \check{b}\right)^c \quad \text{and} \quad f \,\nabla^*\, b = \left(f^c \,\Delta^*\, \check{b}\right)^c. \tag{27}$$

The pair (Δ^*, ∇) *forms an adjunction. Similarly, the pair* (Δ, ∇^*) *is also an adjunction, i.e., for* $f, g, b \in \mathcal{F}(\mathbb{R}^n, \overline{\mathbb{R}})$ *one has*

$$f \,\nabla\, b \leq g \iff f \leq g \,\Delta^*\, b, \tag{28}$$

$$f \,\nabla^*\, b \leq g \iff f \leq g \,\Delta\, b. \tag{29}$$

These relationships are summarized in the following intertwining diagram:

$$\begin{array}{ccc} (f \,\nabla\, b) & \overset{adjoint}{\longleftrightarrow} & (f \,\Delta^*\, b) \\ \updownarrow \textit{dual} & & \updownarrow \textit{dual} \\ (f \,\Delta\, b) & \overset{adjoint}{\longleftrightarrow} & (f \,\nabla^*\, b) \end{array}$$

Proof. Let us derive the adjunction relationship (28) of pair (Δ^*, ∇). First of all, we need to obtain the adjoint operation to the min of two values in a general complete lattice $\mathcal{L}$, whose extreme elements are $\top$ and $\bot$. More precisely, given any triplet $\alpha, t, s \in \mathcal{L}$, as discussed in Section 2, the adjunction involves

$$t \wedge \alpha \leq s \Leftrightarrow t \leq s \wedge^* \alpha,$$

which gives us the corresponding adjoint $\wedge^*$:

$$s \wedge^* \alpha = \sup\{u \in \mathcal{L} \,:\, u \wedge \alpha \leq s\} = \begin{cases} s & \text{if } \alpha > s \\ \top & \text{if } \alpha \leq s \end{cases}$$

Now, for any $f, g, b \in \mathcal{F}(\mathbb{R}^n, \overline{\mathbb{R}})$

$$\begin{aligned} f \,\nabla\, b \leq g &\Leftrightarrow \forall x \in \mathbb{R}^n,\ (f \,\nabla\, b)(x) \leq g(x), \\ &\Leftrightarrow \forall x \in \mathbb{R}^n,\ \sup_{y \in \mathbb{R}^n} \left\{f(y) \wedge b(x-y)\right\} \leq g(x), \\ &\Leftrightarrow \forall x \in \mathbb{R}^n,\ \forall y \in \mathbb{R}^n,\ f(y) \wedge b(x-y) \leq g(x), \\ &\Leftrightarrow \forall x \in \mathbb{R}^n,\ \forall y \in \mathbb{R}^n,\ f(y) \leq g(x) \wedge^* b(x-y), \\ &\Leftrightarrow \forall y \in \mathbb{R}^n,\ f(y) \leq \inf_{x \in \mathbb{R}^n} \left\{g(x) \wedge^* b(x-y)\right\} = (g \,\Delta^*\, b)(y), \\ &\Leftrightarrow f \leq g \,\Delta^*\, b. \end{aligned}$$

The second adjunction can be proved analogously, once the adjoint $\vee^*$ is obtained. We have now

$$t \vee^* \alpha \leq s \Leftrightarrow t \leq s \vee \alpha,$$

such that

$$t \vee^* \alpha = \inf\{u \in \mathcal{L} : t \leq u \vee \alpha\} = \begin{cases} t & \text{if } \alpha < t \\ \perp & \text{if } \alpha \geq t \end{cases}$$

The duality of ∇ and Δ is easily proved

$$\begin{aligned} \left(f^c \nabla \check{b}\right)^c &= \left[\sup_y \left\{f^c(y) \wedge \check{b}(x-y)\right\}\right]^c = \inf_y \{[f^c(y) \wedge b(y-x)]^c\} \\ &= \inf_y \left\{f(y) \vee b^c(y-x)\right\} = f \Delta b. \end{aligned}$$

Similarly, for the duality between ∇^* and Δ^* it is simply required that

$$\left[f^c(y) \wedge^* \check{b}(y-x)\right]^c = f(y) \vee^* b^c(x-y).$$

We find

$$\begin{aligned} \left[f^c(y) \wedge^* \check{b}(y-x)\right]^c &= \left[\begin{cases} f^c(y) & \text{if } b(x-y) > f^c(y) \\ \top & \text{if } b(x-y) \leq f^c(y) \end{cases}\right]^c \\ &= \begin{cases} f(y) & \text{if } b^c(x-y) < f(y) \\ \perp & \text{if } b^c(x-y) \geq f(y) \end{cases} \end{aligned}$$

□

3.3 Commutation with Level Set Processing

We can introduce now the fundamental property of (max, min)-convolutions. We adopt here the convention

$$X \oplus \emptyset = \emptyset \oplus X = \emptyset.$$

Proposition 3. *Let f and b be continuous in $\mathcal{F}(\mathbb{R}^n, \overline{\mathbb{R}})$. Then the four* (max, min)*-convolutions of f by b obey the following commutation rules of level sets with respect to Minkowski sum and subtraction: for all $h \in \overline{\mathbb{R}}$*

$$X_h^+(f \nabla b) = X_h^+(f) \oplus X_h^+(b), \tag{30}$$

$$Y_h^-(f \Delta b) = Y_h^-(f) \oplus Y_h^-(\check{b}^c), \tag{31}$$

$$X_h^+(f \Delta^* b) = X_h^+(f) \ominus X_h^+(b), \tag{32}$$

$$Y_h^-(f \nabla^* b) = Y_h^-(f) \ominus Y_h^-(\check{b}^c). \tag{33}$$

Proof. Let us consider the expression of the supmin convolution (20)

$$X_h^+(f \triangledown b)(x) = \left\{ x \in \mathbb{R}^n ; \bigvee_{y \in \mathbb{R}^n} f(y) \wedge b(x-y) \geq h \right\}.$$

Identifying with terms in Eq. (10), we can write

$$X_h^+(f \triangledown b)(x) = \{ x \in \mathbb{R}^n ; \exists y : [f(y) \wedge b(x-y)] \geq h \}.$$

The expression of $X_h^+(f \triangledown b)$ as function of the $X_h^+(f)$ and $X_h^+(b)$ comes from the relationship (11), which gives

$$\left\{ x \in \mathbb{R}^n : X_h^+(f) \cap X_h^+\left(\check{b}_x\right) \neq \emptyset \right\} = \left\{ x \in \mathbb{R}^n : X_h^+(f) \cap \check{X}_h^+\left(b_x\right) \neq \emptyset \right\},$$

finally from Minkowski addition (2), we obtain

$$X_h^+(f \triangledown b) = X_h^+(f) \oplus X_h^+(b).$$

Now for the infmax convolution $(f \vartriangle b)$, from expression (13), we have

$$\begin{aligned} Y_h^-(f \vartriangle b)(x) &= \left\{ x \in \mathbb{R}^n ; \bigwedge_{y \in \mathbb{R}^n} f(y) \vee b^c(y-x) < h \right\} \\ &= \left\{ x \in \mathbb{R}^n ; \exists y : \left[f(y) \vee \check{b}^c(x-y) \right] < h \right\}, \end{aligned}$$

next, using (12), we get

$$\left\{ x \in \mathbb{R}^n : Y_h^-(f) \cap Y_h^-\left(b_x^c\right) \neq \emptyset \right\} = \left\{ x \in \mathbb{R}^n : Y_h^-(f) \cap \check{Y}_h^-\left(\check{b}_x^c\right) \neq \emptyset \right\},$$

so

$$Y_h^-(f \vartriangle b) = Y_h^-(f) \oplus Y_h^-(\check{b}^c).$$

The expression for the adjoint infmax $(f \vartriangle^* b)$ can be obtained just by adjunction:

$$\begin{aligned} &\sup\left\{ h \in \overline{\mathbb{R}} : x \in X_h^+(f \triangledown b) \right\} \leq g(x) \\ \Leftrightarrow \quad & \cup \{ k \geq h : X_k^+(f \triangledown b) \} \subseteq X_h^+(g) \;\; \forall h \in \overline{\mathbb{R}}, \\ \Leftrightarrow \quad & X_k^+(f \triangledown b) \subseteq X_h^+(g), \;\; \forall h \in \overline{\mathbb{R}}, \; \forall k \geq h, \\ \Leftrightarrow \quad & X_k^+(f) \oplus X_k^+(b) \subseteq X_h^+(g), \;\; \forall h \in \overline{\mathbb{R}}, \; \forall k \geq h, \end{aligned}$$

$$\begin{aligned}
&\Leftrightarrow \quad X_k^+(f) \subseteq X_h^+(g) \ominus X_k^+(b), \ \forall h \in \overline{\mathbb{R}}, \ \forall k \geq h, \\
&\Leftrightarrow \quad X_k^+(f) \subseteq X_h^+(g) \ominus X_k^+(b), \ \forall k \in \overline{\mathbb{R}}, \ \forall h \leq k, \\
&\Leftrightarrow \quad X_k^+(f) \subseteq \cap \left\{h \leq k \, : \, X_h^+(g) \ominus X_k^+(b)\right\}, \ \forall k \in \overline{\mathbb{R}}, \\
&\Leftrightarrow \quad X_k^+(f) \subseteq \cap \left\{h \leq k \, : \, X_h^+(g)\right\} \ominus X_k^+(b), \ \forall k \in \overline{\mathbb{R}}, \\
&\Leftrightarrow \quad X_k^+(f) \subseteq X_k^+(g) \ominus X_k^+(b), \ \forall k \in \overline{\mathbb{R}}, \\
&\Leftrightarrow \quad f(x) \leq \sup \left\{k \in \overline{\mathbb{R}} \, : \, x \in \left(X_k^+(g) \ominus X_k^+(b)\right)\right\},
\end{aligned}$$

and consequently, we have

$$X_k^+\left(g \, \Delta^* \, b\right) = X_k^+(g) \ominus X_k^+(b).$$

By a similar mechanism of adjunction, the expression for the adjoint supmin $(f \, \nabla^* \, b)$ is obtained from $Y_h^-(f \, \Delta \, b) = Y_h^-(f) \oplus Y_h^-(b^c)$.

By the way, using the equivalence $X_h^+(f^c) = X_{-h}^-(f)$, we formulate the level set representation for the infmax convolution by duality by complement from the supmin convolution, we obtain

$$\begin{aligned}
\left(f^c \, \nabla \, \check{b}\right)^c &= \left[\sup\{h \in \mathbb{R} \, : \, X_h^+(f^c) \oplus X_h^+(\check{b})\}\right]^c \\
&= \inf\{-h \in \mathbb{R} \, : \, X_h^+(f^c) \oplus X_h^+(\check{b})\} \\
&= \inf\{-h \in \mathbb{R} \, : \, X_{-h}^-(f) \oplus X_h^+(\check{b})\} \\
&= \inf\{h \in \mathbb{R} \, : \, X_h^-(f) \oplus X_{-h}^+(\check{b})\} \\
&= \inf\{h \in \mathbb{R} \, : \, X_h^-(f) \oplus X_h^-(\check{b}^c)\} = \left(f \, \Delta \, b\right).
\end{aligned}$$

□

This expression on strict lower level sets Y_h^- for $(f \, \Delta \, b)$ is valid for lower level sets X_h^- if $(f \, \Delta \, b)$ is *exact*, in the sense that, for each $x \in \mathrm{dom}^-(f \, \Delta \, b)$, there exists $y \in \mathbb{R}^n$ such that $(f \, \Delta \, b)(x) = f(y) \vee b^c(y - x)$ (i.e., the minimum is attained for any x in the domain) (Seeger & Volle, 1995; Luc & Volle, 1997). In particular, if f and b^c are both LSC quasiconvex functions, $(f \, \Delta \, b)$ and $(f \, \nabla^* \, b)$ are exact, which involves:

$$X_h^-(f \, \Delta \, b) = X_h^-(f) \oplus X_h^-(\check{b}^c), \tag{34}$$

$$X_h^-(f \, \nabla^* \, b) = X_h^-(f) \ominus X_h^-(\check{b}^c). \tag{35}$$

We need for the sequel an alternative formulation of the infmax and adjoint supmin convolution in terms respectively of Minkowski subtraction

$\ominus$ and addition $\oplus$ of level sets. It is simply based on rewriting the infmax convolution using upper level sets:

$$\begin{aligned}(f \vartriangle b)(x) &= \inf\{h \in \mathbb{R} : x \in Y_h^-(f \vartriangle b)\} \\ &= \sup\{h \in \mathbb{R} : x \notin Y_h^-(f \vartriangle b)\} \\ &= \sup\{h \in \mathbb{R} : x \in [Y_h^-(f \vartriangle b)]^c\} \\ &= \sup\left\{h \in \mathbb{R} : x \in \left[Y_h^-(f) \oplus Y_h^-(\check{b}^c)\right]^c\right\} \\ &= \sup\left\{h \in \mathbb{R} : x \in X_h^+(f) \ominus \check{Y}_h^-(\check{b}^c)\right\} \\ &= \sup\{h \in \mathbb{R} : x \in (X_h^+(f) \ominus Y_h^-(b^c))\}. \end{aligned} \tag{36}$$

Analogously, one obtains the following equivalence for the adjoint supmin convolution:

$$\begin{aligned}(f \triangledown^* b)(x) &= \inf\left\{h \in \mathbb{R} : x \in \left(Y_h^-(f) \ominus Y_h^-(\check{b}^c)\right)\right\} \\ &= \sup\{h \in \mathbb{R} : x \in (X_h^+(f) \oplus Y_h^-(b^c))\}. \end{aligned} \tag{37}$$

Therefore, we can write

$$X_h^+(f \vartriangle b) = X_h^+(f) \ominus Y_h^-(b^c), \tag{38}$$

$$X_h^+(f \triangledown^* b) = X_h^+(f) \oplus Y_h^-(b^c). \tag{39}$$

As an immediate consequence of Proposition 3, one gets

$$\begin{aligned}\mathrm{dom}^+(f \triangledown b) &= \mathrm{dom}^+(f) \oplus \mathrm{dom}^+(b), \\ \mathrm{dom}^-(f \vartriangle b) &= \mathrm{dom}^-(f) \oplus \mathrm{dom}^-(b),\end{aligned}$$

where

$$\begin{aligned}\mathrm{dom}^+(\phi) &= \{x \in \mathbb{R}^n : \phi(x) > -\infty\}, \\ \mathrm{dom}^-(\phi) &= \{x \in \mathbb{R}^n : \phi(x) < +\infty\},\end{aligned}$$

stand for the effective domains of $\phi \in \mathcal{F}(\mathbb{R}^n, \overline{\mathbb{R}})$.

Remark. We should clarify that the interest of these expressions of (max, min)-convolutions in terms of Minkowski addition and subtraction is not to suggest an implementation by level set processing followed by stacking. On the contrary, the idea is to show how some morphological operators formulated by level set transforms have a straightforward implementation using (max, min)-convolutions. By the way, the results that we provided are extremely useful to state other properties.

Bibliographic remark. Classically in convex analysis literature (Seeger & Volle, 1995), definition of *level set addition* of f by g is given by $(f \triangle g)(x) = \inf_y \{f(x-y) \vee g(y)\}$ and its symmetric operation as $(f \triangledown g)(x) = \sup_y \{f(x-y) \wedge g(y)\}$ such that $(f \triangledown g) = -((-f) \triangledown (-g))$. This was historically motivated by the fact that the level sum property was pioneered by Rockafellar (1970) as $\{f \triangle g < \alpha\} = \{f < \alpha\} + \{g < \alpha\}$, $\forall \alpha \in \overline{\mathbb{R}}$, where the notation $\{h < \alpha\} = \{x \in \mathbb{R}^n : h(x) < \alpha\}$ is used for the strict lower level set of h and $+$ refers in this context to Minkowski addition.

3.4 Further Properties

Proposition 4. *We have the following properties for the* (max, min)*-convolutions.*

1. *(Increaseness) If* $f(x) \leq g(x)$, $\forall x \in \mathbb{R}^n$, $f, g \in \mathcal{F}(\mathbb{R}^n, \overline{\mathbb{R}})$, *then* $(f \triangledown b)(x) \leq (g \triangledown b)(x)$, $(f \triangle b)(x) \leq (g \triangle b)(x)$, $(f \triangledown^* b)(x) \leq (g \triangledown^* b)(x)$ *and* $(f \triangle^* b)(x) \leq (g \triangle^* b)(x)$, $\forall x \in \mathbb{R}^n$ *and for any* $b \in \mathcal{F}(E, \overline{\mathbb{R}})$.
2. *(Extreme values preservation) Given any function* $f \in \mathcal{F}(\mathbb{R}^n, \overline{\mathbb{R}})$ *and structuring function* $b \in \mathcal{F}(\mathbb{R}^n, \overline{\mathbb{R}})$, *one has*

$$\begin{aligned} \max\left(f \triangledown b\right) &= \left(\max f\right) \wedge \left(\max b\right), \\ \min\left(f \triangle b\right) &= \left(\min f\right) \vee \left(\min b^c\right) = \left(\min f\right) \vee \left(\max b\right), \\ \max\left(f \triangledown^* b\right) &= \max f, \\ \min\left(f \triangle^* b\right) &= \min f. \end{aligned}$$

3. *(Distributivity – commutation with supremum and infimum) Given a structuring function* $b \in \mathcal{F}(\mathbb{R}^n, \overline{\mathbb{R}})$ *and an arbitrary family* $\{f_i\}$, $i \in I$, $f_i \in \mathcal{F}(\mathbb{R}^n, \overline{\mathbb{R}})$, *it follows* $\forall x \in E$ *that*

$$\left(\sup_{i\in I} f_i \triangledown b\right)(x) = \sup_{i\in I}\left(f_i \triangledown b\right)(x); \quad \left(\inf_{i\in I} f_i \triangle b\right)(x) = \inf_{i\in I}\left(f_i \triangle b\right)(x);$$

$$\left(\sup_{i\in I} f_i \triangledown^* b\right)(x) = \sup_{i\in I}\left(f_i \triangledown^* b\right)(x); \quad \left(\inf_{i\in I} f_i \triangle^* b\right)(x) = \inf_{i\in I}\left(f_i \triangle^* b\right)(x).$$

4. *(Associativity – combination of several structuring functions) Given any function* $f \in \mathcal{F}(\mathbb{R}^n, \overline{\mathbb{R}})$ *and any pair of structuring functions* $b_1, b_2 \in \mathcal{F}(\mathbb{R}^n, \overline{\mathbb{R}})$, *one obtains*

$$\begin{aligned} \left(f \triangledown b_1\right) \triangledown b_2 &= f \triangledown \left(b_1 \triangledown b_2\right) = f \triangledown \left(b_2 \triangledown b_1\right), \\ \left(f \triangle b_1\right) \triangle b_2 &= f \triangle \left(b_1 \triangledown b_2\right) = f \triangle \left(b_2 \triangledown b_1\right), \\ \left(f \triangle^* b_1\right) \triangle^* b_2 &= f \triangle^* \left(b_1 \triangledown b_2\right) = f \triangle^* \left(b_2 \triangledown b_1\right), \\ \left(f \triangledown^* b_1\right) \triangledown^* b_2 &= f \triangledown^* \left(b_1 \triangledown b_2\right) = f \triangledown^* \left(b_2 \triangledown b_1\right). \end{aligned}$$

Proof. The increaseness is straightforward since all the operations involved in the (max, min)-convolutions for a fixed b are increasing.

The proof for the extreme value preservation of supmin convolution is as follows:

$$\begin{aligned}\max\left(f \triangledown b\right)(x) &= \sup_{x\in\mathbb{R}^n}\sup_{y,z\in\mathbb{R}^n}\left\{f(y)\wedge b(z),\ \ y+z=x\right\}\\ &= \sup_{y,z\in\mathbb{R}^n}\left\{f(y)\wedge b(z)\right\} = \left(\sup_{y\in\mathbb{R}^n} f(y)\right)\wedge\left(\sup_{z\in\mathbb{R}^n} b(z)\right).\end{aligned}$$

Similarly for $\min\left(f \vartriangle b\right)$. For the adjoint supmin convolution, we have:

$$\begin{aligned}\max\left(f \triangledown^* b\right)(x) &= \sup_{x\in\mathbb{R}^n}\sup_{\substack{y,z\in\mathbb{R}^n\\ y+z=x}}\begin{cases} f(y) & \text{if } b^c(z) < f(y)\\ \perp & \text{if } b^c(z) \geq f(y)\end{cases}\\ &= \sup_{y,z\in\mathbb{R}^n}\begin{cases} f(y) & \text{if } b^c(z) < f(y)\\ \perp & \text{if } b^c(z) \geq f(y)\end{cases}\\ &= \left(\sup_{y,z\in\mathbb{R}^n, b^c(z)<f(y)} f(y)\right)\vee\left(\sup_{y,z\in\mathbb{R}^n, b^c(z)\geq f(y)} \perp\right) = \max f.\end{aligned}$$

Analogously for $\max\left(f \vartriangle^* b\right)$.

We prove the distributivity of supmin convolution and the adjoint supmin convolution. For the two other operators the procedure is similar. One has for the supmin convolution of a sup of functions:

$$\begin{aligned}\left(\sup_{i\in I} f_i \triangledown b\right)(x) &= \sup_{y\in\mathbb{R}^n}\left\{\sup_{i\in I} f_i(y)\wedge b(x-y)\right\}\\ &= \sup_{i\in I}\sup_{y\in\mathbb{R}^n}\left\{f_i(y)\wedge b(x-y)\right\} = \left(f_i \triangledown b\right)(x).\end{aligned}$$

In the case of the adjoint supmin convolution of the sup of function:

$$\left(\sup_{i\in I} f_i \triangledown^* b\right)(x) = \sup_{y\in\mathbb{R}^n}\begin{cases}\sup_{i\in I} f_i(y) & \text{if } b^c(x-y) < \sup_{i\in I} f_i(y)\\ \perp & \text{if } b^c(x-y) \geq \sup_{i\in I} f_i(y)\end{cases}$$

We note the cases where $b^c(x-y) \geq \sup_{i\in I} f_i(y)$ involves $b^c(x-y) \geq f_i(y)$, $\forall i\in I$. Therefore we can write

$$\left(\sup_{i\in I} f_i \triangledown^* b\right)(x) = \sup_{i\in I}\sup_{y\in\mathbb{R}^n}\begin{cases} f_i(y) & \text{if } b^c(x-y) < f_i(y)\\ \perp & \text{if } b^c(x-y) \geq f_i(y)\end{cases} = \sup_{i\in I}\left(f_i \triangledown^* b\right)(x).$$

The associativity of structuring functions is easily obtained using the level set expressions and the following classical results (Matheron, 1975)

$$(X \oplus B_1) \oplus B_2 = X \oplus (B_1 \oplus B_2); \quad (X \ominus B_1) \ominus B_2 = X \ominus (B_1 \oplus B_2).$$

Thus, we have:

$$\begin{aligned}
X_h^+ ((f \triangledown b_1) \triangledown b_2) &= X_h^+ (f \triangledown b_1) \oplus X_h^+ (b_2) \\
&= X_h^+ (f) \oplus X_h^+ (b_1) \oplus X_h^+ (b_2) \\
&= X_h^+ (f) \oplus X_h^+ (b_1 \triangledown b_2).
\end{aligned}$$

$$\begin{aligned}
Y_h^- ((f \vartriangle b_1) \vartriangle b_2) &= Y_h^- \left(f \vartriangle \check{b}_1^c\right) \oplus Y_h^- \left(\check{b}_2^c\right) \\
&= Y_h^- (f) \oplus Y_h^- \left(\check{b}_1^c\right) \oplus Y_h^- \left(\check{b}_2^c\right) \\
&= Y_h^- (f) \oplus \left(Y_{-h}^+ \left(\check{b}_1\right) \oplus Y_-^+ \left(\check{b}_2\right)\right) \\
&= Y_h^- (f) \oplus \check{Y}_{-h}^+ (b_1 \triangledown b_2) \\
&= Y_h^- (f) \oplus \check{Y}_h^- \left((b_1 \triangledown b_2)^c\right).
\end{aligned}$$

$$\begin{aligned}
X_h^+ ((f \vartriangle^* b_1) \vartriangle^* b_2) &= X_h^+ (f \vartriangle^* b_1) \ominus X_h^+ (b_2) \\
&= [X_h^+ (f) \ominus X_h^+ (b_1)] \ominus X_h^+ (b_2) \\
&= X_h^+ (f) \ominus (X_h^+ (b_1) \oplus X_h^+ (b_2)) \\
&= X_h^+ (f) \ominus X_h^+ (b_1 \triangledown b_2).
\end{aligned}$$

$$\begin{aligned}
Y_h^- ((f \triangledown^* b_1) \triangledown^* b_2) &= Y_h^- (f \triangledown^* b_1) \ominus Y_h^- \left(\check{b}_2^c\right) \\
&= \left[Y_h^- (f) \ominus Y_h^- \left(\check{b}_1^c\right)\right] \ominus Y_h^- \left(\check{b}_2^c\right) \\
&= Y_h^- (f) \ominus \left(Y_h^- \left(\check{b}_1^c\right) \oplus Y_h^- \left(\check{b}_2^c\right)\right) \\
&= Y_h^- (f) \ominus \left(\check{Y}_{-h}^+ (b_1) \oplus \check{Y}_{-h}^+ (b_2)\right) \\
&= Y_h^- (f) \ominus \check{Y}_{-h}^+ (b_1 \triangledown b_2) \\
&= Y_h^- (f) \ominus \check{Y}_h^- \left((b_1 \triangledown b_2)^c\right).
\end{aligned}$$

□

Canonic structuring function. The conic structuring function plays a role similar to the multiscale quadratic structuring function (Dorst & van den Boomgaard, 1994; Maragos, 1995; Jackway & Deriche, 1996) in (max, +)-algebra.

Definition 5. The multiscale conic structuring function is defined as the canonic structuring function in (max, min)-convolutions:

$$c_\lambda(x) = -\frac{\|x\|}{\lambda}. \tag{40}$$

In order to justify this canonicity, let us consider the upper level sets of $c_\lambda(x)$. First, we remind that a ball of radius centered at point x is given by the set

$$B_r(x) = \left\{y \in \mathbb{R}^n \,:\, \|x - y\| \leq r\right\}.$$

Proposition 6. *The canonic structuring function in* (max, min)*-convolutions satisfies the semi-group*

$$\left(c_\lambda \bigtriangledown c_\mu\right)(x) = c_{\lambda+\mu}(x). \tag{41}$$

In the case of the L^∞ *metric, a dimension separability is obtained for* $c_\lambda^\infty(x) = -\|x\|_\infty/\lambda$*; i.e., let us denote the coordinates of point as* $x = (x_1, x_2, \cdots, x_n)$ *and by* $c_{\lambda;\,i}(x) = -|x_i|/\lambda$ *the one dimensional conic structuring function, we have*

$$c_\lambda^\infty(x) = \left(c_{\lambda;\,1} \bigtriangledown c_{\lambda;\,2} \cdots \bigtriangledown c_{\lambda;\,n}\right). \tag{42}$$

Proof. We first note that $X_{-h}^{+}(c_\lambda) = B_{\lambda h}$. Second, we remind the Minkowski addition of balls: $B_{r_1} \oplus B_{r_2} = B_{r_1+r_2}$. Therefore, one has

$$X_{-h}^{+}(c_\lambda \bigtriangledown c_\mu) = X_{-h}^{+}(c_\lambda) \oplus X_{-h}^{+}(c_\mu) = B_{\lambda h} \oplus B_{\mu h} = B_{(\lambda+\mu)h}.$$

Dimension separability in L^∞ metric is also a consequence of the Minkowski addition of segments. □

As a consequence of the L^∞ dimension separability, the classical theory of Minkowski decomposition of structuring elements (Serra, 1982), e.g., approximate isotropic structuring elements such as octagons in the square grid or dodecagons in the hexagonal grid using one dimensional structuring elements, can be extended to the case of functions in (max, min)-convolutions.

Concerning this point, we note that the dimension separability of the Euclidean quadratic structuring in the (max, +)-convolutions is a richer property leading to efficient decompositions of separable concave structuring functions (Engbers, Van Den Boomgaard, & Smeulders, 2001).

3.5 Openings, Closings Using (max, min)-Convolutions and Granulometries

The adjointness of the pairs $(\triangle^*, \triangledown)$ and $(\triangle, \triangledown^*)$ involves that from an algebraic viewpoint both the supmin convolution $\triangledown$ and the adjoint supmin convolution $\triangledown^*$ are a dilation; both the infmax convolution $\triangle$ and the adjoint infmax convolution $\triangle^*$ are an erosion. Therefore, their composition naturally yields openings and closings. Let us be more precise.

Definition 7. Given any continuous function $f \in \mathcal{F}(\mathbb{R}^n, \overline{\mathbb{R}})$, the (max, min)-opening and (max, min)-closing of f by the continuous structuring function $b \in \mathcal{F}(\mathbb{R}^n, \overline{\mathbb{R}})$ are respectively given by

$$(f \lozenge b) = ((f \triangle^* b) \triangledown b), \tag{43}$$

and

$$(f \blacklozenge b) = ((f \triangledown^* b) \triangle b), \tag{44}$$

such that their corresponding level sets representations, based on expressions (30), (32), and (31), (33), are given by

$$\begin{aligned} X_h^+(f \lozenge b) &= X_h^+(f \triangle^* b) \oplus X_h^+(b) = [X_h^+(f) \ominus X_h^+(b)] \oplus X_h^+(b) \\ &= X_h^+(f) \circ X_h^+(b), \end{aligned} \tag{45}$$

$$\begin{aligned} Y_h^-(f \blacklozenge b) &= Y_h^-(f \triangledown^* b) \oplus Y_h^-\left(\check{b}^c\right) = \left[Y_h^-(f) \ominus Y_h^-\left(\check{b}^c\right)\right] \oplus Y_h^-\left(\check{b}^c\right) \\ &= Y_h^-(f) \circ Y_h^-\left(\check{b}^c\right). \end{aligned} \tag{46}$$

We note that (max, min)-opening is defined from adjunction $(\triangle^*, \triangledown)$ whereas (max, min)-closing from $(\triangle, \triangledown^*)$. We can also switch roles and to formulate the so-called second family of dual (max, min)-opening and closing as

$$(f \lozenge^* b) = ((f \triangle b) \triangledown^* b), \tag{47}$$

$$(f \blacklozenge^* b) = ((f \triangledown b) \triangle^* b), \tag{48}$$

which has the following equivalent interpretation in terms of level sets:

$$Y_h^-(f \lozenge^* b) = Y_h^-(f) \bullet Y_h^-\left(\check{b}^c\right), \tag{49}$$

$$X_h^+(f \blacklozenge^* b) = X_h^+(f) \bullet X_h^+(b). \tag{50}$$

Besides the duality by complement, classical properties of opening and closing hold in the (max, min) framework.

Proposition 8. *For any $f, b \in \mathcal{F}(\mathbb{R}^n, \overline{\mathbb{R}})$, we have:*

1. *$(f \lozenge b)$ and $(f \blacklozenge b)$ are dual operators*

$$(f \lozenge b) = \left(f^c \blacklozenge \check{b}\right)^c \quad \text{and} \quad (f \blacklozenge b) = \left(f^c \lozenge \check{b}\right)^c; \tag{51}$$

2. *$(f \lozenge b)$ and $(f \blacklozenge b)$ are both increasing operators;*
3. *$(f \lozenge b)$ is anti-extensive and $(f \blacklozenge b)$ extensive with the following ordering relationship:*

$$(f \lozenge b)(x) \leq f(x) \leq (f \blacklozenge b)(x), \ \forall x \in \mathbb{R}^n; \tag{52}$$

4. *idempotency of both operators: $\big((f \lozenge b) \lozenge b\big) = (f \lozenge b)$ and $\big((f \blacklozenge b) \blacklozenge b\big) = (f \blacklozenge b)$.*

Property 1 can be easily proved:

$$\begin{aligned}(f^c \blacklozenge b) &= ((f^c \triangledown^* b) \vartriangle b) = ((f \vartriangle^* \check{b})^c \vartriangle b) = \left(\left(f \vartriangle^* \check{b}\right) \triangledown \check{b}\right)^c \\ &= \left(f \lozenge \check{b}\right)^c.\end{aligned}$$

The other properties are a consequence of the adjunction (Heijmans, 1994).

We note that the second family of (max, min)-opening and closing, $(f \lozenge^* b)$ and $(f \blacklozenge^* b)$, do satisfy the same properties.

Property 4 on idempotency together with the increaseness defines a family of operators called algebraic openings/closings (Serra, 1988; Heijmans, 1994), larger than the one associated to the composition of a pair of dilation and erosion (structural openings). Idempotent and increasing operators are also known as ethmomorphisms (Kiselman, 2007, 2010), in addition, anti-extensivity and extensivity involve that the opening is an anoiktomorphism and the closing a cleistomorphism. One of the most classical results in morphological operators provided us an example of algebraic opening: given a collection of openings $\{\gamma_i\}$, increasing, idempotent and anti-extensive operators for all i, their supremum $\sup_i \gamma_i$ is also an opening (Matheron, 1975). A dual result is obtained for the closing by changing the sup to the inf. The class of openings (resp. closings) is neither closed under infimum (resp. opening) or a generic composition. There is however a semi-group property leading to a scale-space framework for opening/closing operators, known as granulometries. The notion of granulometry in Euclidean morphology is summarized in the following result due to Matheron (1975) and Serra (1988).

Theorem 9. *A parameterized family $\{\gamma_\lambda\}_{\lambda>0}$ of flat openings from $\mathcal{F}(E,\mathcal{T})$ into $\mathcal{F}(E,\mathcal{T})$ is a granulometry (or size distribution) when*

$$\gamma_{\lambda_1}\gamma_{\lambda_2} = \gamma_{\lambda_2}\gamma_{\lambda_1} = \gamma_{\sup(\lambda_1,\lambda_2)}; \quad \lambda_1, \lambda_2 > 0. \tag{53}$$

Condition (53) *is equivalent to both*

$$\begin{aligned} &\gamma_{\lambda_1} \le \gamma_{\lambda_2}; \quad \lambda_1 \ge \lambda_2 > 0, \\ &\mathcal{B}_{\lambda_1} \subset \mathcal{B}_{\lambda_2}; \quad \lambda_1 \ge \lambda_2 > 0, \end{aligned} \tag{54}$$

where $\mathcal{B}_\lambda$ is the invariance domain of the opening at scale λ; i.e., the family of structuring elements Bs such that $B = \gamma_\lambda(B)$ (Serra, 1988).

By duality, we introduce antisize distributions as the families of closings $\{\varphi_\lambda\}_{\lambda>0}$.

Formula (53) shows how translation invariant flat openings are composed and highlights their semi-group structure. The equivalent condition (54) emphasizes the monotonicity of the granulometry with respect to λ: the opening becomes more and more active as λ increases. When dealing with Euclidean spaces, Matheron (1975) introduced the notion of Euclidean granulometry as the size distribution being translationally invariant and compatible with homothetics, i.e.,

$$\gamma_\lambda(f(x)) = \lambda\gamma_1(f(\lambda^{-1}x)), \ \forall\lambda > 0, \tag{55}$$

where $f \in \mathcal{F}(E,\mathcal{T})$ is a Euclidean gray-level image. More precisely, a family of mappings γ_λ is a Euclidean granulometry if and only if there exists a class $\mathcal{B}'$ such that

$$\gamma_\lambda(f) = \bigvee_{B\in\mathcal{B}'}\bigvee_{\mu\ge\lambda}\gamma_{\mu B}(f).$$

Then, we note that condition (55) involves that the domain of invariance $\mathcal{B}_\lambda$ is equal to $\lambda\mathcal{B}$, where $\mathcal{B}$ is the class closed under union, translation and homothetics ≥ 1, which is generated by $\mathcal{B}'$. If we reduce the class $\mathcal{B}'$ to a single element B, the associated size distribution becomes

$$\gamma_\lambda(f) = \bigvee_{\mu\ge\lambda}\gamma_{\mu B}(f).$$

The following key result simplifies the situation. The size distribution by a compact structuring element B is equivalent to $\gamma_\lambda(f) = \gamma_{\lambda B}(f)$ if and only if B is convex (Matheron, 1975).

The extension of the granulometric theory to nonflat structuring functions in the framework of (max, +)-based morphology was deeply studied in Kraus et al. (1993). In particular, it was proved that one can build gray-level Euclidean granulometries with a multiscale structuring function b_λ, i.e., $\gamma_\lambda(f) = \left((f \ominus b_\lambda) \oplus b_\lambda\right)$ if and only if structuring function b_λ has a convex compact domain and is constant there (i.e., flat function). In the terminology of Kraus et al. (1993), (max, +)-granulometries are based on the notion of T-openings.

In the case of (max, min)-openings, we can naturally extend Matheron axiomatic of Euclidean granulometries without the flatness limitation. The transposition of general formulation of granulometries given in Theorem 9 involves that, fixing $b_1, b_2 \in \mathcal{F}(\mathbb{R}^n, \overline{\mathbb{R}})$, we have that

$$(f \lozenge b_1) \leq (f \lozenge b_2) \text{ for every } f \in \mathcal{F}(\mathbb{R}^n, \overline{\mathbb{R}})$$

if and only if $(b_1 \lozenge b_2) = b_1$.

We focus on the interesting case of Euclidean granulometries from (max, min)-openings.

Proposition 10. *Given a structuring function $b_1 \in \mathcal{F}(\mathbb{R}^n, \overline{\mathbb{R}})$ such that all its upper level sets $X_h^+(b_1)$ are convex sets, the family of multi-scale* (max, min)*-openings $\{f \lozenge b_\lambda\}_{\lambda \geq 1}$, where the structuring function at scale λ is given by*

$$b_\lambda(x) = b_1\left(\lambda^{-1}x\right),$$

forms a Euclidean granulometry on any image $f \in \mathcal{F}(\mathbb{R}^n, \overline{\mathbb{R}})$, i.e.,

$$(f \lozenge b_\lambda) = \lambda \star \left(\left(\lambda^{-1} \star f\right) \lozenge b_1\right), \tag{56}$$

which involves compatibility with scaling in the spatial domain, in the sense of Matheron's axiomatic defined as follows

$$\left(\lambda \star f\right)(x) = f\left(\lambda^{-1}x\right), \quad \forall \lambda \geq 1.$$

In addition, we have the following semi-group properties, $\forall \lambda_1, \lambda_2 \geq 1$

$$b_{\lambda_1+\lambda_2}(x) = (b_{\lambda_1} \triangledown b_{\lambda_2})(x), \tag{57}$$

$$\left((f \lozenge b_{\lambda_1}) \lozenge b_{\lambda_2}\right)(x) = \left((f \lozenge b_{\lambda_2}) \lozenge b_{\lambda_1}\right)(x) = \left(f \lozenge b_{\sup(\lambda_1,\lambda_2)}\right)(x). \tag{58}$$

Proof. We need to prove the expression (56). In terms of levels sets, spatial scaling compatibility involves that we have for every $f \in \mathcal{F}(\mathbb{R}^n, \overline{\mathbb{R}})$

$$X_h^+\left(\lambda \star f\right) = \lambda X_h^+\left(f\right), \forall h \in \overline{\mathbb{R}}.$$

Therefore, we have $X_h^+(b_\lambda) = \lambda X_h^+(b_1)$. Convexity of upper level sets of b_1 involves we have the homothetic compatibility of semi-group (57):

$$X_h^+\left(b_{\lambda_1} \triangledown b_{\lambda_2}\right) = \lambda_1 X_h^+(b_1) \oplus \lambda_1 X_h^+(b_2) = (\lambda_1 + \lambda_2)\, X_h^+(b_1) \\ = X_h^+(b_{\lambda_1+\lambda_2}).$$

In addition, one has

$$X_h^+(f) \circ X_h^+(b_\lambda) = X_h^+(f) \circ \lambda X_h^+(b_1).$$

Therefore, the granulometric semi-group (58) is verified at each upper level set:

$$\left(X_h^+(f) \circ X_h^+(b_{\lambda_1})\right) \circ X_h^+(b_{\lambda_2}) = \left(X_h^+(f) \circ X_h^+(b_{\lambda_2})\right) \circ X_h^+(b_{\lambda_1}) \\ = X_h^+(f) \circ X_h^+(b_{\sup(\lambda_1,\lambda_2)}).$$

We have already all the elements for (56):

$$\begin{aligned}
(f \Diamond b_\lambda)(x) &= \sup\left\{h \in \mathbb{R} \,:\, x \in X_h^+\left(f \Diamond b_\lambda\right)\right\} \\
&= \sup\left\{h \in \mathbb{R} \,:\, x \in \left[X_h^+\left(f\right) \circ X_h^+\left(b_\lambda\right)\right]\right\} \\
&= \sup\left\{h \in \mathbb{R} \,:\, x \in \left[X_h^+\left(f\right) \circ \lambda X_h^+\left(b_1\right)\right]\right\} \\
&= \sup\left\{h \in \mathbb{R} \,:\, x \in \left[\lambda X_h^+\left(\lambda^{-1} \star f\right) \circ \lambda X_h^+\left(b_1\right)\right]\right\} \\
&= \sup\left\{h \in \mathbb{R} \,:\, \lambda^{-1} x \in \left[X_h^+\left(\lambda^{-1} \star f\right) \circ X_h^+\left(b_1\right)\right]\right\} \\
&= \sup\left\{h \in \mathbb{R} \,:\, \lambda^{-1} x \in X_h^+\left(\left(\lambda^{-1} \star f\right) \Diamond b_1\right)\right\} \\
&= \left(\left(\lambda^{-1} \star f\right) \Diamond b_1\right)\left(\lambda^{-1} x\right) \\
&= \left[\lambda \star \left(\left(\lambda^{-1} \star f\right) \Diamond b_1\right)\right](x).
\end{aligned}$$

□

A dual result of anti-granulometry is obtained for (max, min) multiscale closings $\left(f \blacklozenge b_\lambda\right)$.

A good candidate of multi-scale isotropic structuring function leading to (max, min) granulometries is based on the canonic structuring function (41), as $b_\lambda(x) = c_\lambda(x) + \alpha$, which is equivalent to

$$b_\lambda(x) = \lambda^{-1} c_1(x) + \alpha, \;\; \lambda \geq 1, \;\alpha > 0.$$

More generally, we can also consider the family of multi-scale functions

$$b_\lambda(x) = -\frac{\|x\|^P}{\lambda^P} + \alpha, \;\; \lambda \geq 1, \;\alpha > 0, \; P > 0.$$

4. HOPF–LAX–OLEINIK FORMULAS FOR HAMILTON–JACOBI EQUATION $u_t \pm H(u, Du) = 0$

Operators formalized in the previous Section are also related to the viscosity solution of a well known family of nonlinear partial differential equations: the Hamilton–Jacobi equation of the form $u_t \pm H(u, Du) = 0$. Hamilton–Jacobi equations are of great importance in physics and optimal control (Clarke and Stern, 1999). Interesting results on Hamilton–Jacobi equations in imaging sciences can be found in Darbon (2015). Before developing the connection with the present work, let us review very briefly the link between another form of Hamilton–Jacobi equation, $u_t \pm H(Du) = 0$, and classical morphological operators viewed as convolutions in $(\max, +)$-algebra.

4.1 Morphological PDE for Classical Dilation and Erosion

Consider the Hamilton–Jacobi equation

$$u_t(x, t) \pm H(x, Du(x, t)) = 0 \text{ in } \mathbb{R}^n \times (0, +\infty), \tag{59}$$

with the initial condition $u(x, 0) = f(x)$ in $\mathbb{R}^n$. Such family of equations usually does not admit classic (i.e., everywhere differentiable) solutions but can be studied in the framework of the theory of viscosity solutions (Crandall, Ishii, & Lions, 1992). It is well known (Bardi & Evans, 1984) that if the Hamiltonian has the properties: (i) $H(x, p) = H(p)$ is convex and (ii) $\lim_{|p|\to+\infty} H(p)/|p| = +\infty$, then the solution of Cauchy problem (59) is given for $+$ and $-$ respectively by the so-called Hopf–Lax–Oleinik formulas:

$$u(x, t) = \inf_{y\in\mathbb{R}^n} \left[f(y) + tH^*\left(\frac{x-y}{t}\right)\right],$$

and

$$u(x, t) = \sup_{y\in\mathbb{R}^n} \left[f(y) - tH^*\left(\frac{x-y}{t}\right)\right],$$

where $H^*(q)$ is the Legendre–Fenchel transform of function $H(p)$, i.e.,

$$H^*(q) = \sup_{p\in\mathbb{R}^n} \left\{ \langle p, q\rangle - H(p)\right\}, \quad q \in \mathbb{R}^n, \tag{60}$$

where $\langle p, q\rangle$ denotes the dot product of p and q.

PDE (59) plays a central role in continuous morphology (Alvarez et al., 1993; Arehart et al., 1993; Brockett & Maragos, 1994; Maragos, 1996). In particular, by taking $H(p) = 1/2\|p\|^2$, such that $H^*(q) = 1/2\|q\|^2$, a kind of canonic morphological PDE is formulated

$$\begin{cases} \frac{\partial u}{\partial t} = \pm\frac{1}{2}\|\nabla u\|^2, & x \in E,\ t > 0 \\ u(x,0) = f(x), & x \in E \end{cases} \tag{61}$$

such that the corresponding viscosity solutions are given by

$$u(x,t) = \sup_{y\in E}\left\{f(y) - \frac{\|x-y\|^2}{2t}\right\} \quad (\text{for } + \text{ sign}), \tag{62}$$

$$u(x,t) = \inf_{y\in E}\left\{f(y) + \frac{\|x-y\|^2}{2t}\right\} \quad (\text{for } - \text{ sign}), \tag{63}$$

which just correspond to a dilation $(f \oplus p_t)(x)$ and an erosion $(f \ominus p_t)(x)$ of image $f(x)$ using the multiscale quadratic (or parabolic) structuring function:

$$p_t(x) = -\frac{\|x\|^2}{2t}. \tag{64}$$

By the way, due to its properties of semi-group, dimension separability and invariance to transform domain (Maragos, 1995; Jackway & Deriche, 1996; van den Boomgaard & Dorst, 1997), the structuring function $p_t(x)$ can be considered as the canonic one in morphology, playing a similar role to the Gaussian kernel in linear filtering.

Other generalized forms of the Hamilton–Jacobi model (59) cover the flat morphology by disks (Maragos, 1996); i.e., $u_t = \pm\|\nabla u\|$, as well as operators with more general P-power concave structuring functions, i.e., $u_t = \pm\|\nabla u\|^P$. For the application of the latter model to adaptive morphology see Diop and Angulo (2015).

4.2 Viscosity Solution of Hamilton–Jacobi Equation with Hamiltonians Containing u and Du

We study now the Hopf–Lax–Oleinik type formulas for Hamilton–Jacobi PDE of form $u_t \pm H(u, Du) = 0$ and its links to convolutions in (max, min)-algebra. The theory of this equation was developed by Barron et al. (1996, 1997). Other interesting results can be found in paper by Alvarez et al. (1999) and the excellent survey paper by Van and Son (2006). The most relevant elements for us can be summarized in the following result.

Proposition 11. *Let us consider the following two Cauchy problems (first-order Hamilton–Jacobi PDEs):*

$$\begin{cases} u_t + H_1(u, Du) = 0, & \text{in } (x,t) \in \mathbb{R}^n \times (0,\infty), \\ u(x,0) = f(x), & \forall x \in \mathbb{R}^n, \end{cases} \tag{65}$$

and

$$\begin{cases} u_t + H_2(u, Du) = 0, & \text{in } (x,t) \in \mathbb{R}^n \times (0,\infty), \\ u(x,0) = g(x), & \forall x \in \mathbb{R}^n, \end{cases} \tag{66}$$

where the initial conditions are functions $f, g : \mathbb{R}^n \times \mathbb{R}$, such that f is an LSC proper function, bounded from below; and g a USC proper function, bounded from above. The Hamiltonians $H_1, H_2 : \mathbb{R} \times \mathbb{R}^n \to \mathbb{R}^n$ are assumed to satisfy the following conditions:

(A1) *$H_1(\gamma, p)$ and $H_2(\gamma, p)$ are continuous;*
(A2) *$H_1(\gamma, p)$ and $H_2(\gamma, p)$ are nondecreasing in $\gamma \in \mathbb{R}$, $\forall p \in \mathbb{R}^n$;*
(A3) *$H_1(\gamma, p)$ is convex and $H_2(\gamma, p)$ is concave in $p \in \mathbb{R}^n$, $\forall \gamma \in \mathbb{R}$;*
(A4) *$H_1(\gamma, p)$ and $H_2(\gamma, p)$ are positively homogeneous of degree 1 in $p \in \mathbb{R}^n$, i.e., $H_1(\gamma, \lambda p) = \lambda H_1(\gamma, p)$, $\forall \lambda \geq 0$.*

The LSC viscosity solution of (65) is given by

$$u(x,y) = \inf_{y \in \mathbb{R}^n} \left[f(y) \vee H_1^{\sharp}\left(\frac{x-y}{t}\right) \right], \tag{67}$$

and the USC viscosity solution of (66) is

$$u(x,y) = \sup_{y \in \mathbb{R}^n} \left[f(y) \wedge H_{2\sharp}\left(\frac{x-y}{t}\right) \right], \tag{68}$$

where the conjugate operators $H^{\sharp}$ and $H_{\sharp}$ are defined as

$$H^{\sharp}(q) = \inf\left\{\gamma \in \mathbb{R} : H(\gamma, p) \geq \langle p, q \rangle, \forall p \in \mathbb{R}^n\right\}, \tag{69}$$
$$H_{\sharp}(q) = \sup\left\{\gamma \in \mathbb{R} : H(\gamma, p) \leq \langle p, q \rangle, \forall p \in \mathbb{R}^n\right\}. \tag{70}$$

Proof. The proof for Cauchy problem (65) corresponds to the one of Theorem 5.5 in Alvarez et al. (1999). Problem (66) is symmetrically obtained by considering the relationship (Van & Son, 2006):

$$\left[-H(-\gamma, -p)\right]^{\sharp}(z) = -\left[H_{\sharp}(\gamma, p)\right](z). \tag{71}$$

□

The simplest case of admissible (A1)–(A4) convex Hamiltonian corresponds to $H(\gamma, p) = \gamma \|p\|$ such that, using Cauchy–Schwartz inequality, one gets

$$H^{\sharp}(q) = \inf \{\gamma \in \mathbb{R} : \gamma \|p\| \geq \langle p, q \rangle\} = \|q\|.$$

The associated concave Hamiltonian is given by $H(\gamma, p) = -\gamma \|p\|$, whose conjugate is also $H_{\sharp}(q) = \|q\|$. Using this case as a starting point, a prototype of PDE in the framework of operators in (max, min)-algebra can be defined.

Definition 12. Given any continuous and bounded function $f : E \to [a, b] \subset \mathbb{R}$, the canonic (Hamilton–Jacobi) PDE in (max, min)-morphology is defined as

$$\begin{cases} \frac{\partial u}{\partial t} = \pm u \|\nabla u\|, & x \in E,\ t > 0 \\ u(x, 0) = f(x), & x \in E \end{cases} \tag{72}$$

and its (unique weak) solutions at scale t are given by

$$u(x, t) = \sup_{y \in E} \left\{ f(y) \wedge \frac{\|x - y\|}{t} \right\} \quad \text{(for + sign)}, \tag{73}$$

$$u(x, t) = \inf_{y \in E} \left\{ f(y) \vee \frac{\|x - y\|}{t} \right\} \quad \text{(for − sign)}. \tag{74}$$

Therefore the viscosity solutions of Cauchy problem (72) are a supmin convolution and an infmax convolution using the conic structuring function $c_\lambda(x)$ given by (77), where the scale parameter is here the time; i.e., $\lambda = t$. More precisely, we note that these solutions

$$\begin{aligned} u(x, t) &= (f \triangledown (-c_t))(x) \quad \text{(for + sign)}, \\ u(x, t) &= (f \vartriangle c_t)(x) \quad \text{(for − sign)}, \end{aligned}$$

are neither dual nor adjoint in the sense of Section 2. Consequently their composition does not lead to opening or closing since for instance:

$$\left((f \triangledown (-c_t)) \vartriangle c_t \right) \neq \left((f \triangledown (-c_t)) \vartriangle^* (-c_t) \right).$$

The model (72) can be generalized to

$$\frac{\partial u}{\partial t} = \pm \alpha u \|\nabla u\|, \quad x \in E,\ t > 0$$

with initial condition $u(x,0) = f(x)$ and $\alpha > 0$, such that we easily see that the corresponding solutions are

$$\begin{aligned} u(x,t) &= (f \mathbin{\triangledown} (-c_{\alpha t}))(x) \quad (\text{for} + \text{sign}), \\ u(x,t) &= (f \mathbin{\vartriangle} c_{\alpha t})(x) \quad (\text{for} - \text{sign}), \end{aligned}$$

or in other words, multiplying u by α involves a scaling in time by α. This principle can be a clue to explore the notion of *spatially adaptive* (max, min)-operators based on using a scale depending on space x, i.e., a model of the form $u_t = \pm\alpha(x)u\|\nabla u\|$.

Let us mention an alternative natural generalization of the canonic PDE which involves raising function u to power P, $P > 0$, given by

$$\frac{\partial u}{\partial t} = \pm u^P \|\nabla u\|, \quad x \in E,\ t > 0,$$

again with initial condition $u(x,0) = f(x)$. Now the corresponding viscosity solutions are

$$u(x,t) = \sup_{y \in E} \left\{ f(y) \wedge \left(-c_{P;t}(x-y)\right) \right\} \quad (\text{for} + \text{sign}), \tag{75}$$

$$u(x,t) = \inf_{y \in E} \left\{ f(y) \vee \left(-c_{P;t}(x-y)\right) \right\} \quad (\text{for} - \text{sign}), \tag{76}$$

where

$$c_{P;t}(x) = -\frac{\|x\|^{\frac{1}{P}}}{t^{\frac{1}{P}}}. \tag{77}$$

We note that $c_{P;t}(x)$ is a concave function if $P \leq 1$ and quasiconcave if $P > 1$ (see next section).

As a final variation of the model of the canonical PDE (72), we can consider the case where the Hamiltonians are replaced by either the pair

$$\left.\begin{aligned} H_1(\gamma,p) &= (\alpha+\gamma)\|p\| \\ H_2(\gamma,p) &= -(\alpha+\gamma)\|p\| \end{aligned}\right\} \Leftrightarrow H_1^\sharp(q) = H_{\sharp 2}(q) = \|q\| - \alpha; \tag{78}$$

or the pair

$$\left.\begin{aligned} H_1(\gamma,p) &= (\alpha-\gamma)\|p\| \\ H_2(\gamma,p) &= -(\alpha-\gamma)\|p\| \end{aligned}\right\} \Leftrightarrow H_1^\sharp(q) = H_{\sharp 2}(q) = \alpha - \|q\|; \tag{79}$$

which involves respectively the structuring functions:

$$c^{\vee}_{\alpha,t}(x) = -\left(\frac{\|x\|}{t} - \alpha\right) \text{ and } c^{\wedge}_{\alpha,t}(x) = -\left(\alpha - \frac{\|x\|}{t}\right).$$

These models are related to the notion of viscous morphology, discussed below.

Another interesting example is the following convex case: $H(\gamma, p) = e^{\gamma}\|p\|$ which satisfies all (A1)–(A4). One computes the conjugate $H^{\sharp}(q)$ as follows: the inequality $H(\gamma, p) = e^{\gamma}\|p\| \geq \langle p, q\rangle$, $\forall p \in \mathbb{R}^n$, holds if and only if $e^{\gamma} \geq \|q\|$, or $\gamma \geq \log\|q\|$. Thus $H^{\sharp}(q) = \log\|q\|$. For the corresponding concave $H(\gamma, p) = -e^{-\gamma}\|p\|$, using relationship (71), we have that $H_{\sharp}(q) = -\log\|q\|$.

Finally, let us bring here the case of a convex–concave Hamiltonian which leads to a curious pair of composed operators. Consider a Cauchy problem of the form

$$\begin{cases} u_t + H(u, Du), & \text{in } (x, t) \in \mathbb{R}^n \times (0, \infty), \\ u(x, 0) = u_0(x), & \forall x \in \mathbb{R}^n, \end{cases}$$

where the Hamiltonian is given by $H(\gamma, p) = H_1(\gamma, p) + H_2(\gamma, p)$, such that H_1 and H_2 hold assumptions (A1)–(A4). In particular $H_1(\gamma, \cdot)$ is convex and $H_2(\gamma, \cdot)$ is concave. As shown in Van and Son (2006), the expected subsolution and supersolution are given by

$$u_-(x, t) = \sup_{z\in\mathbb{R}^n} \inf_{y\in\mathbb{R}^n} \left\{u_0\left(x - t(y + z)\right) \vee H_1^{\sharp}\left(y\right) \wedge H_{2\sharp}\left(z\right)\right\}, \tag{80}$$

$$u_+(x, t) = \inf_{z\in\mathbb{R}^n} \sup_{y\in\mathbb{R}^n} \left\{u_0\left(x - t(y + z)\right) \wedge H_{2\sharp}\left(y\right) \vee H_1^{\sharp}\left(z\right)\right\}. \tag{81}$$

In addition, if u_0 is a bounded and uniformly continuous function, this problem admits a unique viscosity solution $u(x, t)$ such that $u_- \leq u \leq u_+$.

As example of application, we can consider the following Cauchy problem introduced in Van and Son (2006):

$$\begin{cases} \frac{\partial u}{\partial t} = -\|\nabla u\| \sinh u, & x \in E,\ t > 0 \\ u(x, 0) = f(x), & x \in E \end{cases} \tag{82}$$

where $\sinh a$ is the hyperbolic sin function; i.e., $\sinh a = 1/2\left(e^a - e^{-a}\right)$, $\forall a \in \mathbb{R}$. The Hamiltonian $H(\gamma, p) = \|p\| \sinh\gamma$ can be written as $H(\gamma, p) = H_1(\gamma, p) + H_2(\gamma, p) = 1/2e^{\gamma}\|p\| - 1/2e^{-\gamma}\|p\|$, meeting the assumption of sum of convex and concave functions. As we have shown above: $H_1^\sharp(q) = \log\left(2\|q\|\right)$ and $H_{2\sharp}(q) = -\log\left(2\|q\|\right)$. Hence the subsolution and supersolution are derived from (80) and (81); i.e.,

$$u_-(x,t) = \sup_{z\in\mathbb{R}^n}\inf_{y\in\mathbb{R}^n}\left\{f\left(x - t(y+x)\right) \vee \log\left(2\|y\|\right) \wedge -\log\left(2\|z\|\right)\right\}, \quad (83)$$

$$u_+(x,t) = \inf_{z\in\mathbb{R}^n}\sup_{y\in\mathbb{R}^n}\left\{f\left(x - t(y+x)\right) \wedge -\log\left(2\|y\|\right) \vee \log\left(2\|z\|\right)\right\}. \quad (84)$$

Using our notation, we have

$$u_-(x,t) = \left(\left(f \vartriangle l_t\right) \triangledown l_t\right)(x),$$
$$u_+(x,t) = \left(\left(f \triangledown l_t\right) \vartriangle l_t\right)(x),$$

with $l_t(x) = -\log\left(2\|x\|t^{-1}\right)$. We can expect that, even if the solution of (82) is neither an opening nor a closing, it will produce a rather symmetric regularization of a bounded real valued function f. That corresponds to a kind of self-dual filtering with respect to 0.

5. NONLINEAR ANALYSIS USING OPERATORS $(\vartriangle, \triangledown)$

Besides their relationships with first-order Hamilton–Jacobi PDE, (max, min)-convolutions are powerful tools for other areas of nonlinear mathematics.

5.1 Quasi-Concavity

Minkowski addition and subtraction behave particularly well for convex sets (Matheron, 1975). If X is convex, $X \ominus B$ is convex for any B. If X and B are convex, $X \oplus B$ is convex. The interest of (max, +)-convolutions (i.e., classical dilation $(f \oplus b)$ and erosion $(f \ominus b)$ of functions) in convex analysis is known from the origin of this area of mathematics (Moreau, 1970; Rockafellar, 1970). Convolutions in (max, min)-algebra play a similar, but less known, role in the case of a more general class of functions: the quasiconvex functions.

Let us start by a recall on quasiconvexity/quasiconcavity (Avriel, Diewert, Schaible, & Zang, 1988). A function $f \in \mathcal{F}(C, \overline{\mathbb{R}})$ defined on a convex set $C \subset \mathbb{R}^n$, $C \in \mathcal{C}$, is said to be quasiconvex if, for every $x, y \in \mathbb{R}^n$ and $\lambda \in (0,1)$, one has

$$f\left(\lambda x + (1-\lambda)y\right) \leq f(x) \vee f(y). \quad (85)$$

On the contrary, $f \in \mathcal{F}(C, \overline{\mathbb{R}})$ is said to be quasiconcave if

$$f\left(\lambda x + (1-\lambda)y\right) \geq f(x) \wedge f(y). \tag{86}$$

Note that f is quasiconvex if and only if $-f$ (or $M-f$ in the case of functions bounded in [0, 1]) is quasiconcave. Roughly speaking, we can say that any "unimodal function" is quasiconcave.

An alternative way of defining quasiconvexity and quasiconvexity involves level sets. A function $f \in \mathcal{F}(C, \overline{\mathbb{R}})$ is quasiconvex if every lower level set of f is convex, i.e., $X_h^-(f) \in \mathcal{C}$. A function $f \in \mathcal{F}(C, \overline{\mathbb{R}})$ is quasiconcave if every upper level set of f is convex $X_h^+(f) \in \mathcal{C}$. Using formulas (14) and (15), we note that the definition of quasiconvexity is valid by replacing the convexity of lower level set by the convexity of strict lower level set.

The notion of quasiconcavity is weaker than the notion of concavity, in the sense that every concave function is quasiconcave. Similarly, every convex function is quasiconvex.

Note also that a concave function can be a quasiconvex function, e.g., $x \mapsto \log(x)$ is concave, and it is quasiconvex.

We can now introduce the following immediate result.

Proposition 13. *Let $b : C \to \overline{\mathbb{R}}$ be a quasiconcave function defined on a convex set $C \subset \mathbb{R}^n$, $C \in \mathcal{C}$.*
- *If $g \in \mathcal{F}(C, \overline{\mathbb{R}})$ is a quasiconvex function, then the infmax convolution $(g \,\Delta\, b)$ is quasiconvex.*
- *If $f \in \mathcal{F}(C, \overline{\mathbb{R}})$ is a quasiconcave function, then the supmin convolution $(f \,\nabla\, b)$ is quasiconcave.*

Proof. The proof is straightforward from the formulation of supmin and infsup convolutions in terms of respectively Minkowski sum of upper level sets and lower level sets; i.e., expressions (30) and (31), together with the property of convexity preservation of Minkowski addition $X \oplus B$ if both X and B are convex. □

In addition, if we observe the expressions (32) and (33) of the adjoint operators, which involve Minkowski subtraction, we have a straightforward more general result:
- If $g \in \mathcal{F}(C, \overline{\mathbb{R}})$ is a quasiconvex function, then the adjoint infmax convolution $(g \,\Delta^*\, b)$ is quasiconvex for any structuring function $b \in \mathcal{F}(C, \overline{\mathbb{R}})$.

- If $f \in \mathcal{F}(C, \overline{\mathbb{R}})$ is a quasiconcave function, then the adjoint supmin convolution $(f \bigtriangledown^* b)$ is quasiconcave for any structuring function $b \in \mathcal{F}(C, \overline{\mathbb{R}})$.

As a consequence, for the product operators, openings and closings, we also have nice properties of quasiconcavity/convexity preservation.

Proposition 14. *Let $b : C \to \overline{\mathbb{R}}$ be a quasiconcave function defined on a convex set $C \subset \mathbb{R}^n$, $C \in \mathcal{C}$.*

- *If $g \in \mathcal{F}(C, \overline{\mathbb{R}})$ is a quasiconcave function, then the* (max, min)*-opening $(g \Diamond b)$ is quasiconcave.*
- *If $f \in \mathcal{F}(C, \overline{\mathbb{R}})$ is a quasiconvex function, then the* (max, min)*-closing $(f \blacklozenge b)$ is quasiconvex.*

Proof. In both cases the result is trivial from the expressions of (max, min)-opening and closing in terms of level sets. For instance, one has for the closing:

$$(f \blacklozenge b) = \left((f \bigtriangledown^* b) \bigtriangleup b\right).$$

So, if b is quasiconcave then $\check{b}^c$ is quasiconvex, then as f is quasiconvex, the operator $(f \bigtriangledown^* b)$ is quasiconvex, so $\left((f \bigtriangledown^* b) \bigtriangleup b\right)$ too. □

The study of connectedness of level sets under (max, min)-convolutions is also straightforward from the properties of connectedness of Minkowski addition; i.e., if X and B are connected, $X \oplus B$ is connected (Matheron, 1975). The following assertions hold (Luc & Volle, 1997): (i) if f and b have connected strict lower (resp. strict upper) sets then so does $(f \bigtriangleup b)$ (resp. $(f \bigtriangledown b)$); (ii) if f and b have connected lower (resp. upper) sets then so does $(f \bigtriangleup b)$ (resp. $(f \bigtriangledown b)$). Other issues related to connectedness, as the existence of nonlocal minima of $(f \bigtriangleup b)(x)$ or nonlocal maxima of $(f \bigtriangledown b)(x)$ as well as some properties on continuity of level set maps from (max, min)-convolutions have been also studied in Luc and Volle (1997).

5.2 Lipschitz Approximation

In many mathematical areas, from nonlinear analysis to optimization theory, it is important to construct Lipschitz approximation of a given function. We need first to recall the notion of Lipschitzian function. A function $\phi : \mathbb{R}^n \to \mathbb{R}$ is said to be Lipschitz with constant $L \in \mathbb{R}_+$ if

$$|\phi(x) - \phi(y)| \leq L\|x - y\|, \quad \forall x, y \in \mathbb{R}^n.$$

In this respect, (max, min)-convolutions has a fruitful role to play (Seeger & Volle, 1995).

Let us use the notion of proper function f as the a function taking values in the extended real number line such that $f(x) < +\infty$ for at least one x and $f(x) > -\infty$ for every x.

Proposition 15. *Let the structuring function $b : \mathbb{R}^n \to \mathbb{R}$ be majorized from above and L-Lipschitz.*

- *If $f : \mathbb{R}^n \to \mathbb{R} \cup \{-\infty\}$ is a proper function, then the supmin convolution $(f \triangledown b)$ is L-Lipschitz.*
- *If $g : \mathbb{R}^n \to \mathbb{R} \cup \{+\infty\}$ is a proper function, then the infmax convolution $(g \triangle b)$ is L-Lipschitz.*

Proof. We provide the proof for the supmin convolution; the case of infmax is obtained similarly. First, the operator can be written in the form

$$(f \triangledown b)(x) = \sup \left\{ \phi_y(x) \ : \ y \in \mathrm{dom}^- f \right\},$$

where

$$\phi_y(x) = f(y) \vee b(x - y).$$

Obviously, since $b(x)$ is L-Lipschitz, for each $y \in \mathrm{dom}^- f$, the function $\phi_y : \mathbb{R}^n \to \mathbb{R}$ is also L-Lipschitz. Indeed, for all x and z in $\mathbb{R}^n$, we can write

$$\begin{aligned} |\phi_y(x) - \phi_y(z)| &= |f(y) \vee b(x - y) - f(y) \vee b(z - y)| \\ &\leq |b(x - y) - b(z - y)| \\ &\leq L\|x - z\|. \end{aligned}$$

This is equivalent to say that

$$\phi_y(z) - L\|x - z\| \leq \phi_y(x) \leq \phi_y(z) + L\|x - z\|,$$

and then, taking the supremum with respect to $y \in \mathrm{dom}^- f$, one gets

$$(f \triangledown b)(z) - L\|x - z\| \leq (f \triangledown b)(x) \leq (f \triangledown b)(z) + L\|x - z\|$$

which is equivalent

$$|(f \triangledown b)(x) - (f \triangledown b)(z)| \leq L\|x - z\|.$$

□

Roughly speaking, this basic result asserts that the supmin and infmax convolutions of f and g by b inherits the Lipschitzian property of b, no matter how bad the functions f and g are.

A well-known case of Lipschitzian function is the conic structuring function:

$$c_\lambda(x) = -\lambda^{-1}\|x\|, \quad \lambda > 0,$$

which is just a λ^{-1}-Lipschitz function. By using our canonic multiscale structuring function c_λ relevant properties of regularization are obtained.

Proposition 16. *Let f be a nonpositive proper function and g be a nonnegative proper function.*

- *(Lipschitzian approximation) The supmin convolution $f^{[\lambda]}(x) = (f \triangledown c_\lambda)(x)$ and the infmax convolution $g_{[\lambda]}(x) = (g \vartriangle c_\lambda)(x)$ are λ^{-1}-Lipschitz.*
- *(Convergence to envelopes) The lower envelope of the family $\{f^{[\lambda]} : \lambda > 0\}$ converges monotonically downward to the upper-semicontinuous hull $\mathrm{cl}^+ f$ of f, i.e.,*

$$(\mathrm{cl}^+ f)(x) = \inf_{\lambda>0} f^{[\lambda]}(x), \quad \forall x \in \mathbb{R}^n.$$

The upper envelope of the family $\{g_{[\lambda]} : \lambda > 0\}$ converges monotonically upward to the lower-semicontinuous hull $\mathrm{cl}^- g$ of g, i.e.,

$$(\mathrm{cl}^- g)(x) = \sup_{\lambda>0} g^{[\lambda]}(x), \quad \forall x \in \mathbb{R}^n.$$

- *(Local extrema preservation) For $f(x_0) \in]-\infty, 0[$, one has the equivalence*

$$f^{[\lambda]}(x_0) = f(x_0) \quad \Longleftrightarrow \quad x_0 \textit{ maximizes } f \textit{ over } \overset{\circ}{B}_{\lambda f(x_0)}(x_0).$$

For $g(x_0) \in]0, +\infty, [$, one has the equivalence

$$g_{[\lambda]}(x_0) = g(x_0) \quad \Longleftrightarrow \quad x_0 \textit{ minimizes } g \textit{ over } \overset{\circ}{B}_{\lambda g(x_0)}(x_0)$$

where $\overset{\circ}{B}_r(x)$ is the open ball centered at x and radius $r > 0$.

Proof. Lipschitzian approximation is a particular case of previous proposition. The proof of convergence of infmax convolution corresponds to that of Proposition 3.2 in Seeger and Volle (1995). For the supmin convolution the same result is obtained by duality.

Let us show how local minima preservation holds, which is based on the proof of Proposition 3.3 in Seeger and Volle (1995). On the one hand, recall that one has always the inequality $g_{[\lambda]}(x_0) \leq g(x_0)$. On the other hand,

$$g(x_0) \leq g(y) \vee \lambda^{-1}\|x_0 - y\|, \quad \forall y \in \mathbb{R}^n,$$

which is equivalent to

$$g(x_0) \leq g(y) \;\; \text{whenever} \;\; \lambda^{-1}\|x_0 - y\| < g(x_0).$$

Therefore, $g_{[\lambda]}(x_0) = g(x_0)$ if and only if $g(y) \geq g(x_0)$ for all $y \in \overset{\circ}{B}_{\lambda g(x_0)}(x_0)$.

The local maxima preservation is obtained analogously, since we have

$$\begin{aligned} f(x_0) &\geq f(y) \wedge \lambda^{-1}\|x_0 - y\|, \quad \forall y \in \mathbb{R}^n, \;\Rightarrow \\ f(x_0) &\geq f(y) \text{ whenever } \lambda^{-1}\|x_0 - y\| < f(x_0) \end{aligned}$$

which together with inequality $f^{[\lambda]}(x_0) \geq f(x_0)$, involves that $f^{[\lambda]}(x_0) = f(x_0)$ if and only if $f(y) \leq f(x_0)$ for all $y \in \overset{\circ}{B}_{\lambda f(x_0)}(x_0)$. □

Regularization effects of supmin and infmax convolutions by the conic structuring functions can be generalized to more general one-parameter kernels. For instance, for $(g \,\Delta\, k_\lambda)$, a kernel $k_\lambda : \mathbb{R}_+ \to \overline{\mathbb{R}}$ of type $k_\lambda\left(\lambda^{-1}\|x\|\right)$ such that $k(\cdot)$ is nondecreasing, $k(0) \leq \max g$ and $\min g < \max k$; e.g., $k_\lambda(x) = \log(\lambda^{-1}\|x\|)$. See main generalized results in Penot and Zălinescu (2001).

Bibliographic remark. The use of standard dilation and erosion for convex functions, i.e., (max, +)-convolutions, for regularization is well known. In particular, the Moreau–Yosida transform of a convex function g, which consists in an erosion (infimal convolution) of g with a parabolic structuring function, produces a Lipschitz continuously differentiable approximation of the function. Note that for the case of a quasiconvex function it works only in the 1-dimensional case.

If we replace the convexity of g by either a condition of boundness on bounded sets or a quadratic minorization property, one can obtain approximation and regularization of these arbitrary functions g in Hilbert spaces by the Lasry–Lions method (Lasry & Lions, 1986; Attouch & Aze, 1993), which consists in a quadratic erosion followed by a quadratic dilation. This technique can be also applied to functions on a Riemannian manifold of nonpositive curvature (Angulo & Velasco-Forero, 2014).

The generalization of the (max, min)-regularization considered here to a counterpart of the Lasry–Lions framework as well as the extension to curved spaces will be the object of ongoing research.

5.3 A Transform in (max, min)-Convolution

The Fourier transform of the classical convolution in $(+, \times)$ algebra of two functions is the product of their corresponding transforms. Similarly, in the context of convex analysis, the Legendre–Fenchel transform of the $(\max, +)$ inf-convolution coincides with the sum of the corresponding Legendre–Fenchel transforms.

We recall the Legendre–Fenchel transform of a function $\phi : E \to \mathbb{R} \cup \{+\infty\}$ is the function $\phi^* : E^* \to \mathbb{R} \cup \{-\infty\}$ given by

$$\phi^*(w) = \sup_{x \in E} \{\langle w, x\rangle - \phi(x)\}, \quad \forall w \in E^*, \tag{87}$$

where E and its Fenchel conjugate space E^* are subsets of $\mathbb{R}^n$. In that what follows, the symbol Λ stands for the 2D simplex

$$\Lambda = \left\{(\lambda_1, \lambda_2) \in \mathbb{R}^2 \,:\, \lambda_1 + \lambda_2 = 1,\ \lambda_1 \geq 0,\ \lambda_2 \geq 0\right\}.$$

We need to introduce also the notion of right multiplication of ϕ^* by a finite scalar $\lambda \geq 0$ as follows

$$\left(\phi^*\lambda\right)(w) = \begin{cases} \lambda\phi^*\left(\lambda^{-1}w\right) & \text{if } \lambda > 0 \\ \sup_{x \in \mathrm{dom}^- \phi} \langle w, x\rangle & \text{if } \lambda = 0 \end{cases}$$

Using these ingredients, relationship between Legendre–Fenchel transform and (max, min)-convolutions was considered in Seeger and Volle (1995). We consider in particular the case for the infmax convolution. Let $f : E \to \mathbb{R} \cup \{+\infty\}$ be a proper convex function and $b : E \to \mathbb{R} \cup \{-\infty\}$ be a proper concave function. The Legendre–Fenchel transform of $(f \Delta b)$ is given by

$$\left(f \,\Delta\, b\right)^*(w) = \inf_{(\lambda_1, \lambda_2) \in \Lambda} \left(f^*\lambda_1 + (\check{b}^c)^*\lambda_2\right)(w), \quad \forall w \in E^*. \tag{88}$$

Even if the formula (88) seems *compact*, it presents from our viewpoint a limited interest since, on the one hand, it is related to a representation of the infmax convolution as

$$(f \,\Delta\, b)(x) = \inf_{(u,v) \in M} \sup_{(\lambda_1, \lambda_2) \in \Lambda} \left\{\lambda_1 f(u) + \lambda_2 \check{b}^c(v)\right\},$$

where

$$M=\left\{(u,v)\in \mathrm{dom}^{-} f \times \mathrm{dom}^{-} \check{b}^{c} : u+v=x\right\}.$$

On the other hand, it is limited to convex functions.

Research on another transform intrinsically adapted to (max, min)-convolutions have been conducted in two independent approaches. Gondran (1996) introduced the so-called infmaxaffine transform; Volle (1998), inspired from the results on Hamilton–Jacobi PDE developed by Barron et al. (1996, 1997) (see previous section), introduced quasiconvex conjugate transform. We adopt here the framework developed by Volle and consequently proof of all results are given in Volle (1998). We have nevertheless adapted the notation and terminology to our context.

Support function. The notion of support function plays a central role in the transform domain considered here. We recall that the support function of a nonempty set $X\subset\mathbb{R}^n$ is defined by

$$i_X^*(y)=\sup_{x\in X}\langle x,y\rangle,\quad \forall y\in\mathbb{R}^n, \tag{89}$$

with the convention $\sup\emptyset=-\infty$. We note that the support function of X is just the Legendre–Fenchel transform of the indicator function i_X of X:

$$i_X(x)=\begin{cases}0 & \text{if } x\in X\\ +\infty & \text{if } x\in X^c\end{cases}$$

As examples, we can mention: (i) the support function of a singleton $X=\{x\}$ is $i_X^*(\lambda y)=\langle x,y\rangle$; (ii) the support function of the Euclidean unit ball $X=B_1$ is $i_X^*(\lambda y)=\|x\|$.

The support function is positively homogeneous, i.e., $i_X^*(\lambda y)=\lambda i_X^*(y)$ $\lambda>0$, $\forall y\in\mathbb{R}^n$ and subadditive, i.e., $i_X^*(y+z)\le i_X^*(y)+i_X^*(z)$, $\forall y,z\in\mathbb{R}^n$. It follows that i_X^* is an LSC convex function.

The support functions of a scaled or translated set are closely related to the original set X: (i) $i_{\lambda X}(y)=\lambda i_X(y)$, $\lambda>0$, $y\in\mathbb{R}^n$; (ii) $i_{X+w}(y)=i_X(y)+\langle x,w\rangle$, $y,w\in\mathbb{R}^n$. The latter generalizes to one of the fundamental properties of the support function: the additivity with respect to the Minkowski addition, i.e., for all $y\in\mathbb{R}^n$ one has

$$i_{X\oplus B}^*(y)=i_X^*(y)\,+.\,i_B^*(y), \tag{90}$$

where $(+\infty)\,+.\,(-\infty)=(-\infty)\,+.\,(+\infty)=-\infty$.

α- and β-conjugates. The motivation of the conjugate domain in (max, min) algebra is founded on the representation of infmax convolution as the Minkowski addition of lower level sets. In fact, the corresponding representation is just based on the support function of each lower level set.

Definition 17. To each extended real-valued function $f : \mathbb{R}^n \to \overline{\mathbb{R}}$, let us associate the α-conjugate function $f^\alpha : \mathbb{R}^n \times \mathbb{R} \to \overline{\mathbb{R}}$ defined by

$$f^\alpha(y, h) = \sup_{f(x)<h} \langle x, y \rangle = i^*_{Y_h^-(f)}. \tag{91}$$

For any $y \in \mathbb{R}^n$, $f^\alpha(y, \cdot)$ is LSC and nondecreasing. Moreover, for any family $(f_i)_{i\in I}$, $f_i \in \mathcal{F}(\mathbb{R}^n, \overline{\mathbb{R}})$, one has

$$\left(\inf_{i\in I} f_i\right)^\alpha = \sup_{i\in I} f_i^\alpha.$$

This α-conjugation is closely related to the infsup convolution (Volle, 1998, Theorem 3.1):

Proposition 18. *Given any pair of functions $f, b \in \mathcal{F}(\mathbb{R}^n, \overline{\mathbb{R}})$, one always has*

$$\left(f \bigtriangleup \check{b}^c\right)^\alpha = f^\alpha \dotplus b^\alpha. \tag{92}$$

This result is an easy consequence of relation (31) and the property (90):

$$\begin{aligned} (f \bigtriangleup b)^\alpha (y, h) &= i^*_{Y_h^-(f \bigtriangleup b)}(y) = i^*_{Y_h^-(f)\oplus Y_h^-(\check{b}^c)} = i^*_{Y_h^-(f)}(y) \dotplus i^*_{Y_h^-\left(\check{b}^c\right)}(y) \\ &= f^\alpha(y, h) \dotplus \left(\check{b}^c\right)^\alpha (y, h). \end{aligned}$$

Let us define now a kind of dual (pseudo-inverse) transform.

Definition 19. The β-conjugate of any function $\phi : \mathbb{R}^n \times \mathbb{R} \to \overline{\mathbb{R}}$ is the extended real-valued function $\phi^\beta : \mathbb{R}^n \to \overline{\mathbb{R}}$ defined as

$$\phi^\beta(x) = \sup_{y\in\mathbb{R}^n} \sup\left\{h \in \mathbb{R} : \langle x, y \rangle > \phi(y, h)\right\}. \tag{93}$$

We also have the property

$$\left(\inf_{i\in I} \phi_i\right)^\beta = \sup_{i\in I} \phi_i^\beta$$

for any family $(\phi_i)_{i\in I}$, $\phi_i \in \mathcal{F}(\mathbb{R}^n \times \mathbb{R}, \overline{\mathbb{R}})$.

The α and β conjugates are related by the following Galois correspondence

$$f^{\alpha} \leq \phi \Leftrightarrow f \geq \phi^{\beta} \tag{94}$$

for any pair of functions $f \in \mathcal{F}(\mathbb{R}^n, \overline{\mathbb{R}})$ and $\phi \in \mathcal{F}(\mathbb{R}^n \times \mathbb{R}, \overline{\mathbb{R}})$.

We consider the composite operators, called the biconjugates

$$\begin{array}{rcl} f \in \mathcal{F}(\mathbb{R}^n, \overline{\mathbb{R}}) & \mapsto & f^{\alpha\beta} = (f^{\alpha})^{\beta} \in \mathcal{F}(\mathbb{R}^n, \overline{\mathbb{R}}), \\ \phi \in \mathcal{F}(\mathbb{R}^n \times \mathbb{R}, \overline{\mathbb{R}}) & \mapsto & \phi^{\beta\alpha} = (\phi^{\beta})^{\alpha} \in \mathcal{F}(\mathbb{R}^n \times \mathbb{R}, \overline{\mathbb{R}}). \end{array}$$

The biconjugates are

- increasing: $f \leq g \Rightarrow f^{\alpha\beta} \leq g^{\alpha\beta}$; $\phi \leq \psi \Rightarrow \phi^{\beta\alpha} \leq \psi^{\beta\alpha}$;
- anti-extensive: $f^{\alpha\beta} \leq f$; $\phi^{\beta\alpha} \leq \phi$;
- idempotent: $(f^{\alpha\beta})^{\alpha\beta} = f^{\alpha\beta}$; $(\phi^{\beta\alpha})^{\beta\alpha} = \phi^{\beta\alpha}$.

The biconjugate of $f \in \mathcal{F}(\mathbb{R}^n, \overline{\mathbb{R}})$, resp. ϕ, is nothing but the greatest regular minorant of f, resp. $\phi \in \mathcal{F}(\mathbb{R}^n \times \mathbb{R}, \overline{\mathbb{R}})$:

$$\begin{array}{rcl} f^{\alpha\beta} & = & \sup\left\{g \in \mathcal{F}(\mathbb{R}^n, \overline{\mathbb{R}}) \,:\, g^{\alpha\beta} = g \text{ and } g \leq f\right\}, \\ \phi^{\beta\alpha} & = & \sup\left\{\psi \in \mathcal{F}(\mathbb{R}^n, \overline{\mathbb{R}}) \,:\, \psi^{\beta\alpha} = \psi \text{ and } \psi \leq \phi\right\}. \end{array}$$

All the above properties are mere consequences of the theory of Galois correspondences.

We can introduce now one of the important results which describes the classes of regular functions in such conjugate domains (Volle, 1998, Theorem 3.4).

Proposition 20. *Let $f : \mathbb{R}^n \to \overline{\mathbb{R}}$ be an extended real-valued function. Then*

$$f = f^{\alpha\beta}$$

if and only if f is an LSC quasiconvex function.

From the previous results, we observe that for any pair of functions $f, b \in \mathcal{F}(\mathbb{R}^n, \overline{\mathbb{R}})$, $(f^{\alpha} \dot{+} b^{\alpha})^{\beta}$ is just the LSC quasiconvex hull of the infimax convolution $f \,\triangle\, \check{b}^c$. In the case when f and b are quasiconvex, $f \,\triangle\, \check{b}^c$ is quasiconvex too so that $(f^{\alpha} \dot{+} b^{\alpha})^{\beta}$ is then the LSC hull of $f \,\triangle\, \check{b}^c$. In order to have the explicit equivalence $f \,\triangle\, \check{b}^c = (f^{\alpha} \dot{+} b^{\alpha})^{\beta}$, f and b should be quasiconvex, compact and satisfy some other minor technical constraints, see (Volle, 1998, Theorem 4.2).

Applications. Let us come back to the first-order Hamilton–Jacobi PDE discussed in Proposition 11. We remind that the LSC viscosity solution with convex Hamiltonian $H(\gamma, p)$, $p \in \mathbb{R}^n$, $\gamma \in \mathbb{R}$, involves the conjugate Hamiltonian (69), given by

$$H^\sharp(q) = \inf\left\{\gamma \in \mathbb{R} : H(\gamma, p) \geq \langle p, q\rangle, \forall p \in \mathbb{R}^n\right\}.$$

As $H(\cdot, p)$ is nondecreasing, we note that the function $H^\sharp(q)$ above is just the β-conjugate of Hamiltonian $H(\gamma, p)$, namely:

$$H^\sharp(q) = \left(H(p,\gamma)\right)^\beta(q) = \sup_{p\in\mathbb{R}^n} \sup\left\{\gamma \in \mathbb{R} : H(p,\gamma) < \langle p, q\rangle\right\}.$$

As we have discussed in Section 4, for the case $H(p, \gamma) = \gamma\|p\|$, we have $H^\beta(q) = \|p\|$. Therefore, α and β conjugates can be applied to solve some Hamilton–Jacobi PDEs.

We conclude this part by describing the conjugates of an important class of quasiconvex functions, namely the radial quasiconvex functions, which generalizes the previous case. Let us consider that we work in a general Banach space E equipped with a norm $\|\cdot\|$ and denote by $\|\cdot\|_*$ the dual norm:

$$\|\cdot\|_*(y) = \max_{\|x\|=1} \langle x, y\rangle,$$

for any y in the topological dual F of E. We remind that if $p, q \in [1, +\infty]$ satisfy $1/p + 1/q = 1$, then the L^p and L^q norms are dual to each other. In particular, the Euclidean norm is self-dual ($p = q = 2$). To any nondecreasing extended real-valued function $a : [0, +\infty[\to \overline{\mathbb{R}}$ is associated the quasiconvex function $f = a \circ \|\cdot\|$, where $\circ$ denotes the composition of functions. Setting

$$\tilde{a}(r) = \begin{cases} a(0) & \text{for } r \in]-\infty, 0[\\ a(r) & \text{for } r \in [0, +\infty[\end{cases}$$

we get a nondecreasing function $\tilde{a} : \mathbb{R} \to \overline{\mathbb{R}}$ such that $\tilde{a}(0) = \inf \tilde{a}$. We will use also the notion of LSC quasi-inverse of a given by

$$a^e(s) = \sup\{r \in \mathbb{R} : a(r) < s\} = \inf\{r \in \mathbb{R} : a(r) \geq s\}.$$

The α-conjugate of a radial quasiconvex function can be written as the product of a nondecreasing function by the dual norm. More precisely, we have the following result (Volle, 1998, Propositions 4.10 and 4.11):

Proposition 21. *Let a and $\tilde{a}$ be as above. Then, one has*

$$(a \circ \|\cdot\|)^{\alpha}(y,h) = \tilde{a}^{e}(h)\|y\|_{*}, \quad \forall y \in F, h \in \mathbb{R}.$$

Let $\phi : F \times \mathbb{R} \to \overline{\mathbb{R}}$ be defined by $\phi(y,h) = b(h)\|y\|_{*}$ with $b : \mathbb{R} \to \overline{\mathbb{R}}$ nondecreasing, then

$$\phi^{\beta}(x) = b^{e}(\|x\|), \quad \forall x \in E.$$

Finally, we have the following result on biconjugates (Volle, 1998, Corollaries 4.12 and 4.13):

Proposition 22. *Let $f = a \circ \|\cdot\|$ with function $a : [0, +\infty[\to \overline{\mathbb{R}}$ nondecreasing. Then $f^{\alpha\beta} = \bar{a} \circ \|\cdot\|$, where $\bar{a}$ is the LSC hull of a. Let $\phi : F \times \mathbb{R} \to \overline{\mathbb{R}}$, $\phi(y,h) = b(h)\|y\|_{*}$ with $b : \mathbb{R} \to \overline{\mathbb{R}}$ nondecreasing, the $\phi^{\beta\alpha} = \bar{b}(h)\|y\|_{*}$.*

Let us conclude by saying that the theory of α, β-conjugates and in particular, the considered case of radial quasiconvex functions, will be applied in the future to identify the interesting cases of structuring functions and their composition rules in the framework of (max, min)-convolutions.

Bibliographic remark. In the state-of-the-art of mathematical morphology, the notion of Legendre–Fenchel transform was rediscovered and generalized to any function, either convex or nonconvex: it is named the slope transform (Dorst & van den Boomgaard, 1994; Maragos, 1995).

Support function defined on convex sets in the plane was used by Schmitt (1996) to formulate Minkowski addition of sets represented by its boundary.

6. UBIQUITY OF (max, min)-CONVOLUTIONS IN MATHEMATICAL MORPHOLOGY

Our aim in this section is to review some previous approaches of "unconventional" dilation and erosion (in the sense that they cannot be related to (max, +)-convolutions paradigm) which fit naturally in a formulation as (max, min)-convolutions.

The two first cases are conventional and rather trivial, however, we consider that links to the other frameworks deserves some discussion.

6.1 Distance Function

The distance function is a powerful tool in image processing. It is based on endowing space E with a particular norm $\|\cdot\|$. Then for any nonempty subset $X \subset E$, the distance function of set X, $d_X : E \to \mathbb{R}_+$, is defined as

$$d_X(x) = \inf_{y \in X^c} \|x - y\|, \quad x \in E.$$

The distance function is just a particular case of the infmax convolution (Seeger & Volle, 1995). Indeed, if one introduces the *zero indicator function* $\mathbb{O}_X$ of set X:

$$\mathbb{O}_X(x) = \begin{cases} 0 & \text{if } x \in X \\ -\infty & \text{if } x \in X^c \end{cases}$$

then, using the unit conic structuring function $c_1(x) = -\|x\|$, one can write:

$$d_X(x) = (-\mathbb{O}_{X^c} \Delta\, c_1)(x). \tag{95}$$

The proof is straightforward:

$$\begin{aligned} (-\mathbb{O}_{X^c} \Delta\, c_1)(x) &= \inf_{y \in E} \{-\mathbb{O}_{X^c}(y) \vee \|x - y\|\} = \begin{cases} \inf_{y \in E} \|x - y\| & \text{if } y \in X^c \\ +\infty & \text{if } y \in X \end{cases} \\ &= \inf_{y \in X^c} \|x - y\|. \end{aligned}$$

By changing the norm in the conic unit structuring function, other distance functions can be obtained. Nevertheless, this infmax convolution is not the most efficient way to compute the distance function.

6.2 Flat Morphological Operators Using Indicator Functions

The binary case of morphological operators is solved by simply using the standard *indicator function* $\mathbb{1}_X$ of a set $X \subset E$:

$$\mathbb{1}_X(x) = \begin{cases} 1 & \text{if } x \in X \\ 0 & \text{if } x \in X^c \end{cases}$$

Then, given a set X and a structuring element B, we have

$$(\mathbb{1}_X \nabla \mathbb{1}_B) = \mathbb{1}_{X \oplus B}, \tag{96}$$

$$\left(\mathbb{1}_X \Delta^* \mathbb{1}_B\right) = \mathbb{1}_{X \ominus B}. \tag{97}$$

This equivalence can be proved, without using upper level set representation, as follows:

$$(\mathbb{1}_X \nabla \mathbb{1}_B)(x) = 1 \quad \Leftrightarrow \quad \exists y \in E,\ \left(\mathbb{1}_X(y) \wedge \mathbb{1}_B(x - y)\right) = 1,$$

$$\begin{aligned}
&\Leftrightarrow \quad \mathbb{1}_X(y) = 1 \text{ and } \mathbb{1}_B(x-y) = 1,\\
&\Leftrightarrow \quad y \in X \text{ and } x - y \in B,\\
&\Leftrightarrow \quad x \in X \oplus B.
\end{aligned}$$

Proof of the second equality can be carried out analogously.

More generally, the case of flat dilation and erosion for functions is also easily obtained by using the *infinity indicator function* ∞_X of set X:

$$\infty_X(x) = \begin{cases} +\infty & \text{if } x \in X \\ -\infty & \text{if } x \in X^c \end{cases}$$

In the case of nonnegative bounded functions, $+\infty$ is replaced by M and $-\infty$ by 0. Then, for any function f in $\mathcal{F}(E, \overline{\mathbb{R}})$ or in $\mathcal{F}(E, [0, M])$, we have:

$$\left(f \triangledown \infty_B\right) = \left(f \oplus B\right), \tag{98}$$
$$\left(f \vartriangle^* \infty_B\right) = \left(f \ominus B\right). \tag{99}$$

The proof is trivial using the representation of the supmin and adjoint infmax convolutions in terms of upper level sets and taking into account that $X_h^+(\infty_B) = B$, $\forall h$; i.e.,

$$\begin{aligned}
X_h^+\left(f \triangledown \infty_B\right) &= X_h^+(f) \oplus X_h^+(\infty_B) = X_h^+(f) \oplus B,\\
X_h^+\left(f \vartriangle^* \infty_B\right) &= X_h^+(f) \ominus X_h^+(\infty_B) = X_h^+(f) \ominus B.
\end{aligned}$$

In addition, we also have that

$$\begin{aligned}
\left(f \vartriangle \infty_B\right) &= \inf_{y \in E} \left\{f(y) \vee \left(-\check{\infty}_B\right)\right\} = \left(f \ominus B\right), &(100)\\
&\Rightarrow \left(f \vartriangle^* \infty_B\right) = \left(f \vartriangle \infty_B\right),\\
\left(f \triangledown^* \infty_B\right) &= \sup_{y \in E} \left\{f(y) \vee^* (-\infty_B)\right\} = \left(f \oplus B\right), &(101)\\
&\Rightarrow \left(f \triangledown^* \infty_B\right) = \left(f \triangledown \infty_B\right),
\end{aligned}$$

and therefore for this case of flat structuring elements, which correspond to cylinders as equivalent structuring functions, the dual and adjoint operators play the same role.

6.3 Links with Fuzzy Morphology

The state-of-the-art on morphological operators based on fuzzy logic is very extensive. We refer to papers by Nachtegael and Kerre (2001) and by

Bloch (2009) which present an excellent survey of the existing literature on fuzzy morphology. The latter brightly discusses the role of duality with respect to complement vs. adjunction in fuzzy approaches. Results on fuzzy morphology discussed here are mainly based on Deng and Heijmans (2002). A complementary deeper insight from the algebraic viewpoint, in particular for representation theorems, can be found in a paper by Maragos (2005).

In fuzzy logic, the two basic (Boolean) logic operators, the conjunction $C(s,t) = s \wedge t$ and the implication $I(s,t) = s \Rightarrow t$ $(= \neg s \vee t)$, are extended from the Boolean domain $\{0,1\} \times \{0,1\}$ to the rectangle $[0,1] \times [0,1]$. A fuzzy conjunction is a mapping from $[0,1] \times [0,1]$ into $[0,1]$ which is increasing in both arguments and satisfies $C(0,0) = C(1,0) = C(0,1) = 0$ and $C(1,1) = 1$. A fuzzy implication is decreasing in the first argument, increasing in the second one and satisfies $I(0,0) = I(0,1) = I(1,1) = 1$ and $I(1,0) = 0$.

Given a fuzzy set μ, the dilation and erosion by a fuzzy structuring element ν are then defined as (Deng & Heijmans, 2002):

$$\begin{aligned} \delta_{\nu,C}(\mu)(x) &= \sup_y \left\{ C\left(\nu(x-y), \mu(y)\right)\right\}, && (102) \\ \varepsilon_{\nu,C}(\mu)(x) &= \inf_y \left\{ I\left(\nu(y-x), \mu(y)\right)\right\}. && (103) \end{aligned}$$

As shown in Deng and Heijmans (2002), (I, C) is an adjunction if and only if $(\varepsilon_{\nu,C}, \delta_{\nu,C})$ is an adjunction. Therefore, fuzzy opening and closing derived from the composition of these operators have all required properties, whatever the choice of the adjunction (I, C). Note however that some properties such as the iteration, i.e.,

$$\delta_{\nu_2,C}\left(\delta_{\nu_1,C}(\mu)\right) = \delta_{\delta_{\nu_1,C}(\nu_2),C}(\mu); \;\; \varepsilon_{\nu_2,C}\left(\varepsilon_{\nu_1,C}(\mu)\right) = \varepsilon_{\delta_{\nu_1,C}(\nu_2),C}(\mu),$$

require C to be associative and commutative.

Two particular cases of conjunction and adjoint implication widely used in fuzzy logic are the Gödel–Brower:

$$\begin{aligned} C_{GB}(a,t) &= \min(a,t), && (104) \\ I_{GB}(a,t) &= \begin{cases} s, & s < a \\ 1, & s \geq a \end{cases} && (105) \end{aligned}$$

and the Kleen–Dienes:

$$C_{KD}(a,t) = \begin{cases} 0, & t \leq 1-a \\ t, & t > 1-a \end{cases} \qquad (106)$$

$$I_{KD}(a,s) \quad = \quad \max(1-a,s). \tag{107}$$

It is consequently straightforward to see that the four operators that we have defined in Section 2 are just fuzzy dilations and erosions when they are applied to fuzzy sets (i.e., functions valued in [0, 1]):

$$\begin{array}{ccc}
\delta_{\nu,C_{GB}}(\mu)(x) = (\mu \bigtriangledown \nu)(x) & \stackrel{\text{adjoint}}{\longleftrightarrow} & \varepsilon_{\nu,C_{GB}}(\mu)(x) = \left(\mu \bigtriangleup^* \nu\right)(x) \\
\updownarrow \text{dual} & & \updownarrow \text{dual} \\
\varepsilon_{\nu,C_{KD}}(\mu)(x) = (\mu \bigtriangleup \nu)(x) & \stackrel{\text{adjoint}}{\longleftrightarrow} & \delta_{\nu,C_{KD}}(\mu)(x) = \left(\mu \bigtriangledown^* \nu\right)(x).
\end{array}$$

Let us conclude clarifying that fuzzy logic is much more general than the pairs of conjunction/adjunction considered here and consequently one cannot simply reduce fuzzy morphology to (max, min)-convolutions.

6.4 Links with Viscous Morphology

Theory and practice of morphological (flat) viscous operators was introduced by Vachier and Meyer (2005, 2007). The PDE formulation of these operators was done by Maragos and Vachier (2008). The seminal idea was inspired from a physical model of image flooding by a viscous fluid (Meyer, 1993).

As we have discussed at the beginning of the paper, a fundamental property of flat operators applied to numerical functions is the fact that they can be computed by applying the corresponding set operator to the upper level sets, followed by stacking the processed upper level sets. In other terms, the same set operator, that depends on a structuring element, acts equally for all upper level sets $X_h^+(f)$: that can be interpreted as a similar behavior on all the structures whatever their intensity h. The idea of viscous operators is to apply a different scale (i.e., size) of structuring element at each upper level set. This principle can be seen now as an operator which locally adapts its activity with respect to the intensity.

Let us formalize their definition according to Maragos and Vachier (2008). For the sake of simplicity, let us consider a nonnegative bounded continuous (or discrete) function $f : E \to [0, M]$. Viscous operators have been formulated as isotropic transforms, that is based on the use of balls B_λ as structuring elements. In this framework, the (isotropic) intensity-adaptive dilation and erosion of function f are defined as

$$\delta_{\lambda(h)}(f) \quad = \quad \sup\left\{h \in [0,M] \,:\, x \in \left(X_h^+(f) \oplus B_{\lambda(h)}\right)\right\},$$

$$\varepsilon_{\lambda(h)}(f) \quad = \quad \inf\left\{h \in [0, M] \,:\, x \in \left(X_h^+(f) \ominus B_{\lambda(h)}\right)\right\},$$

where $\lambda(h) : [0, M] \to [0, M]$ is the scaling function with respect to intensity h. Two types of viscosity (linear) functions have been proposed:

$$\lambda_\wedge(h) \quad = \quad \lambda_0 \frac{M-h}{M} \quad \text{(negative slope)},$$
$$\lambda_\vee(h) \quad = \quad \lambda_0 \frac{h}{M} \quad \text{(positive slope)}.$$

Note that, using $\lambda_\wedge(h)$, points of lowest intensity, $\lambda_\wedge(0) = \lambda_0$, are strongly dilated (resp. eroded) while points of highest intensity, $\lambda_\wedge(M) = 0$, are left unchanged. Opposite effect are obtained for $\lambda_\vee(h)$. In the sequel, one set $\lambda_0 = M$; i.e., $\lambda_\wedge(h) = M - h$ and $\lambda_\vee(h) = h$, such that $\lambda_\wedge$ and $\lambda_\vee$ are dual functions.

Using intensity-adaptive operators and the two viscosity functions, two pairs of viscous dilation and erosion are defined for a given function f:

$$\delta_\wedge^{\text{visc}}(f) = \delta_{\lambda_\wedge(h)}(f) = \sup\left\{h \in [0, M] \,:\, x \in \left(X_h^+(f) \oplus B_{M-h}\right)\right\}, \tag{108}$$
$$\varepsilon_\wedge^{\text{visc}}(f) = \varepsilon_{\lambda_\wedge(h)}(f) = \sup\left\{h \in [0, M] \,:\, x \in \left(X_h^+(f) \ominus B_{M-h}\right)\right\}, \tag{109}$$

and

$$\delta_\vee^{\text{visc}}(f) = \delta_{\lambda_\vee(h)}(f) = \sup\left\{h \in [0, M] \,:\, x \in \left(X_h^+(f) \oplus B_h\right)\right\}, \tag{110}$$
$$\varepsilon_\vee^{\text{visc}}(f) = \varepsilon_{\lambda_\vee(h)}(f) = \sup\left\{h \in [0, M] \,:\, x \in \left(X_h^+(f) \ominus B_h\right)\right\}, \tag{111}$$

such that $\left(\varepsilon_\wedge^{\text{visc}}, \delta_\wedge^{\text{visc}}\right)$ and $\left(\varepsilon_\vee^{\text{visc}}, \delta_\vee^{\text{visc}}\right)$ form two adjunctions. The pairs $\left(\varepsilon_\vee^{\text{visc}}, \delta_\wedge^{\text{visc}}\right)$ and $\left(\varepsilon_\wedge^{\text{visc}}, \delta_\vee^{\text{visc}}\right)$ are dual by complement. Consequently, their compositions produce two viscous openings and two closings: $\gamma_\wedge^{\text{visc}}(f) = \delta_\wedge^{\text{visc}}\left(\varepsilon_\wedge^{\text{visc}}(f)\right)$, $\gamma_\vee^{\text{visc}}(f) = \delta_\vee^{\text{visc}}\left(\varepsilon_\vee^{\text{visc}}(f)\right)$, $\varphi_\wedge^{\text{visc}}(f) = \varepsilon_\wedge^{\text{visc}}\left(\delta_\wedge^{\text{visc}}(f)\right)$ and $\varphi_\vee^{\text{visc}}(f) = \varepsilon_\vee^{\text{visc}}\left(\delta_\vee^{\text{visc}}(f)\right)$.

If we introduce the following structuring function:

$$v(x) = \begin{cases} M - \|x\| & \text{if } \|x\| \leq M \\ 0 & \text{if } \|x\| > M \end{cases}$$

such that its complement structuring function is $v^c(x) = \|x\|$ if $\|x\| \leq M$ and M if $\|x\| > M$, the previous viscous dilations and erosions (108)–(111) can be rewritten using the (max, min)-convolution (respectively expressions (30), (32), (38), (39)):

$$\delta_\wedge^{\text{visc}}(f) \quad = \quad \sup\left\{h \in [0, M] \,:\, x \in \left(X_h^+(f) \oplus X_h^+(v)\right)\right\} = \left(f \triangledown v\right),$$

$$\begin{aligned}
\varepsilon_{\wedge}^{\text{visc}}(f) &= \sup\left\{h \in [0,M] : x \in \left(X_h^+(f) \ominus X_h^+(v)\right)\right\} = \left(f \,\Delta^*\, v\right), \\
\delta_{\vee}^{\text{visc}}(f) &= \sup\left\{h \in [0,M] : x \in \left(X_h^+(f) \oplus Y_h^-(v^c)\right)\right\} = \left(f \,\nabla^*\, v\right), \\
\varepsilon_{\vee}^{\text{visc}}(f) &= \sup\left\{h \in [0,M] : x \in \left(X_h^+(f) \ominus Y_h^-(v^c)\right)\right\} = \left(f \,\Delta\, v\right).
\end{aligned}$$

The link between the corresponding viscous opening and closing and our (max, min)-opening and closing is trivial. In fact, using the results from Proposition 3, adjointness of two pairs of operators is obvious. Concerning the duality by complement, for instance, between $\delta_{\wedge}^{\text{visc}}(f)$ and $\varepsilon_{\vee}^{\text{visc}}(f)$, we have, on the one hand,

$$\begin{aligned}
\left(\delta_{\wedge}^{\text{visc}}(f^c)\right)^c &= \left[\sup\left\{h \in [0,M] : x \in \left(X_h^+(f^c) \oplus X_h^+(v)\right)\right\}\right]^c \\
&= \inf\left\{(M-h) \in [0,M] : x \in \left(X_h^+(f^c) \oplus X_h^+(v)\right)\right\} \\
&= \inf\left\{(M-h) \in [0,M] : x \in \left(X_h^+(M-f) \oplus X_h^+(v)\right)\right\} \\
&= \inf\left\{h \in [0,M] : x \in \left(X_{M-h}^+(M-f) \oplus X_{M-h}^+(v)\right)\right\} \\
&= \inf\left\{h \in [0,M] : x \in \left(X_h^-(f) \oplus X_h^-(v^c)\right)\right\}.
\end{aligned}$$

On the other hand, we have

$$\begin{aligned}
\varepsilon_{\vee}^{\text{visc}}(f) &= \sup\left\{h \in [0,M] : x \in \left(X_h^+(f) \ominus Y_h^-(v^c)\right)\right\} \\
&= \inf\left\{h \in [0,M] : x \notin \left(X_h^-(f) \ominus Y_h^-(v^c)\right)\right\} \\
&= \inf\left\{h \in [0,M] : x \in \left(Y_h^-(f) \oplus Y_h^-(v^c)\right)\right\}.
\end{aligned}$$

Therefore, being precise, we have duality only when $\varepsilon_{\vee}^{\text{visc}}(f)$ is exact, see (34). The difficulty to prove the duality by complement of viscous operators without the equivalence (36) was pointed out by Meyer (2008).

In addition to the operator framework, a PDE formulation of viscous dilation and erosion was introduced in Maragos and Vachier (2008). The two corresponding continuous models are:

$$\begin{cases} \frac{\partial u}{\partial t} = \pm(\max f - u)\|\nabla u\|, & x \in E, \ t > 0 \\ u(x,0) = f(x), & x \in E \end{cases} \tag{112}$$

and

$$\begin{cases} \frac{\partial u}{\partial t} = \pm(u - \min f)\|\nabla u\|, & x \in E, \ t > 0 \\ u(x,0) = f(x), & x \in E \end{cases} \tag{113}$$

This couple of PDEs are particular cases of the Hamilton–Jacobi models discussed in Section 4. More precisely, it corresponds to the case of the

Hamiltonians given in expressions (78) and (79). We can therefore conclude that the solution $u(x,t)$ for $+$ sign of model (112) is equivalent to viscous dilation $\delta_{\wedge}^{\text{visc}}(f)$, but for $-$ sign it is not exactly equivalent to the viscous erosion $\varepsilon_{\wedge}^{\text{visc}}(f)$. In our terminology, the latter is a case of adjoint infmax convolution while the solution for $-$ sign is an infmax convolution with the complemented structuring function. Similarly, solution $u(x,t)$ of model (113) is equivalent to viscous erosion $\varepsilon_{\vee}^{\text{visc}}(f)$ for $-$ sign and once again, for $+$ sign a supmin convolution is obtained.

As a conclusion of this part on viscous operators, let us clarify that when using our formulation of such operators it becomes evident that viscous openings and closings do not require any implementation using upper level set decompositions and then stacking. They can be implemented straightforward as (max, min)-convolutions, which notably reduces the computation load.

6.5 Links with Boolean Random Function Characterization

There is a theory of random set modeling intimately related to morphological image processing. This theory was introduced by Matheron (1975) and yields a sound family of probabilistic models for microstructures (dealing with variability, heterogeneity, multi-scaling, etc.). It has been widely used for modeling and predicting the average macroscopic response of random media from their microstructure. Morphological operators are used to test and to select appropriate models (i.e., estimate their parameters) from images and to simulate new images following the model. The set theory was later generalized to random functions. Our aim here is just to point out how (max, min)-convolutions are related to random function characterization in the framework of this theory.

Boolean random set characterization. First of all, we briefly recall the framework for Boolean random sets. Let us assume that we work in the Euclidean space $\mathbb{R}^n$. A Boolean Random Closed Set (or Boolean RACS) $X \subset \mathbb{R}^n$ of parameters (θ, A') is obtained as follows (Matheron, 1975). Consider, on the one hand, a realization of a Poisson point process of intensity $\theta \geq 0$; i.e., the Poisson points x_i are the germs. It means that the number of points in a Borel set B, denoted $N(B)$, follows a Poisson distribution of parameter $\theta\mu_n(B)$ (i.e., mean proportional to the Lebesgue measure of B):

$$P\left\{N(B) = k\right\} = e^{-\theta\mu_n(B)}\frac{(\theta\mu_n(B))^k}{k!},$$

where $\mu_n(B)$ is the Lebesgue measure of B. Consider, on the other hand, a family of compact closed random sets A' i.i.d. located at the origin; i.e., A' is the primary grain. Then, implant at each germ a realization of the primary grain and take the union for all the points; i.e.,

$$X = \bigcup_{x_i} A'_{x_i}.$$

Any shape can be used for A': convex or nonconvex, connected or not connected.

Given a RACS X, let us introduce the *Choquet capacity* $T_X(K)$, defined on the compact sets $K \in \mathcal{K}$ as the probability that the compact K hits the set X, i.e.,

$$T_X(K) = P\{K \cap X \neq \emptyset\} = 1 - P\{K \subset X^c\} = 1 - Q_X(K).$$

The functional T has the following properties: (i) T is bounded with $0 \leq T \leq 1$ and $T(\emptyset) = 0$; (ii) T is increasing in the sense $T(K) \leq T(K \cup K')$, $K, K' \in \mathcal{K}$; (iii) T is upper-semicontinuous. In fact, T acts as a distribution function of a random variable. If X is a stationary RACS, then its Choquet capacity is shift invariant, i.e., $T_X(K_h) = T_X(K)$. There is a fundamental result, named the *Choquet–Matheron–Kendall theorem*, saying that a RACS is characterized (described completely) by the probabilities $Q_X(K)$ as K spans the class of the compact sets $\mathcal{K}$. From a practical viewpoint, it is important to note that the experimental estimation of $T_X(K)$ and $Q_X(K)$ from images is done using realizations of K and morphological transforms; e.g., dilation or erosion:

$$\begin{aligned} T_X(K_x) &= P\{K_x \cap X \neq \emptyset\} = P\left\{x \in X \oplus \check{K}\right\}; \\ Q_X(K_x) &= P\{K_x \subset X^c\} = P\{x \in X^c \ominus K\}. \end{aligned}$$

These expressions can be simplified in the case of a Boolean RACS. Let X be a Boolean RACS of Poisson density θ and primary grain A'. The Choquet capacity of X for any $K \in \mathcal{K}$ is given by (Matheron, 1975)

$$T_X(K) = 1 - Q_X(K) = 1 - e^{-\theta \mathbb{E}\left\{\mu_n\left(A' \oplus \check{K}\right)\right\}}.$$

Thus the number of primary grains hit by K follows a Poisson distribution with average $\theta \mathbb{E}\left\{\mu_n\left(A' \oplus \check{K}\right)\right\}$. Now taking different K, such as a singleton $K = \{x\}$, a couple of points $K = \left\{x, x + d\right\}$, etc. the moments of

$Q(K)$ are obtained. The empirical values are used to fit the parameters of the model. In addition, if the primary grain is a convex set, using Steiner formula, the expressions of $\mathbb{E}\left\{\mu_n\left(A' \oplus \lambda\check{K}\right)\right\}$ are polynomials in λ. These computations are used to test the validity of the Boolean model assumption.

Choquet capacity of random functions. The Boolean random functions are a generalization of the Boolean RACS. The basic idea of Boolean random functions was introduced by Jeulin and Jeulin (1981) for modeling rough surfaces; this model has been extensively studied and generalized by Serra (1989) and by Préteux and Schmitt (1988). We follow here the results formulated by Jeulin (1991); for a more recent development, see Jeulin (2014).

The functional framework involves to generalize the RACS X by means of an upper semicontinuous function $Z(x)$ and the compact test set K by a lower semicontinuous test function $\tau(x)$. Then, as for the case of random sets, the basic mathematical tool to deal with random functions is the Choquet capacity.

Let $Z(x) : \mathbb{R}^n \to \mathbb{R} \cup \{-\infty\}$ be a USC random function, almost surely bounded and with compact (closed) upper level sets. The test function $\tau(x) : K \to \mathbb{R} \cup \{+\infty\}$ is defined on a compact support $K \in \mathcal{K}$ and it is LSC. The Choquet capacity $T_Z(\tau)$ defined over the set of test functions is now given by

$$T_Z(\tau) = P\{x \in D_Z(\tau)\} = 1 - Q_Z(\tau) = 1 - P\{x \in [D_Z(\tau)]^c\},$$

where $D_Z(\tau)$ is the set

$$D_Z(\tau) = \left\{\exists y \in \mathbb{R}^n \,:\, Z(y) \geq \tau(x-y)\right\}$$

such that the complementary of the event $D_Z(\tau)$ is given by the following random set

$$[D_Z(\tau)]^c = \left\{x \,:\, Z(x+y) < \tau(y),\ \forall y \in K\right\}.$$

By convention it is considered that $\tau(x) = +\infty$ for $x \notin K$. Thus the functional $T_Z(\tau)$ gives the probability that the deterministic function τ "hits" the hypograph of the random function Z, which is a closed set. We should notice that by the lower semicontinuity of τ, its epigraph is a closed set.

For the points x which satisfy $Z(x+y) < \tau(y)$, it holds: $Y_h^-(\tau) \subset \left[X_h^+(Z_x)\right]^c$, $\forall h \in \mathbb{R}$. Thus one gets

$$[D_Z(\tau)]^c \quad = \quad \left\{x : \forall h \in \mathbb{R},\ x \in \left[X_h^+(Z_x)\right]^c \ominus Y_h^-(\tau)\right\}$$

$$= \bigcap_{h \in \mathbb{R}} Y_h^-(Z_x) \ominus Y_h^-(\tau). \tag{114}$$

The dual by complement is given by

$$D_Z(\tau) = \bigcup_{h \in \mathbb{R}} X_h^+(Z_x) \oplus \check{Y}_h^-(\tau). \tag{115}$$

As for the case of random sets, we are only interested in stationary functions; i.e., $T_Z(\tau)$ is invariant to translation of τ.

In order to relate these functionals with (max, min)-convolutions, we only need to remind from (37) that

$$(f \triangledown^* b)(x) \quad = \quad \sup \left\{ h \in \overline{\mathbb{R}} : X_h^+(f) \oplus Y_h^-(b^c) \right\}.$$

Thus we can now write the Choquet capacity $T_Z(\tau)$ using the adjoint supmin convolution:

$$T_Z(\tau) = P\left\{ x \in \bigcup_{h \in \mathbb{R}} X_h^+(Z) \oplus \check{X}_h^-(\tau) \right\} = P\left\{ x \in \bigcup_{h \in \mathbb{R}} X_h^+\left(Z \triangledown^* \check{\tau}^c\right) \right\}. \tag{116}$$

In addition, if we assume that the random function Z is a nonnegative function, and we denote by $[Z]_+$ the support set of $Z(x) > 0$, then we can write

$$T_Z(\tau) \quad = \quad P\left\{ x \in \left[Z \triangledown^* \check{\tau}^c \right]_+ \right\}. \tag{117}$$

It is not surprising to discover that the dilation of X by K in the RACS Choquet capacity is generalized to the adjoint supmin of Z by the τ in the functional Choquet capacity: we only need to note how the adjoint supmin is defined. The complement of τ is due to the definition of $\triangledown^*$ as the adjoint of $\triangle$.

Boolean random functions. The particular case of the Boolean random functions naturally involves the construction of a random function over a Poisson point process. We focus on the case of the so-called Boolean islands (Serra, 1989). Namely, three steps are needed.

1. A Poisson point process in $\mathbb{R}^n$ of intensity θ.
2. At each point of the process x_i, we set up a nonnegative random function $f_i(x)$, called the primary random function, drawn independently from a distribution f'; i.e., $f_i \sim f'$.

3. The Boolean random function $Z(x)$ is obtained as

$$Z(x) = \bigvee_{x_i} f_i(x - x_i). \tag{118}$$

Obviously, the set $X_h^+(Z)$ is a Boolean random set with primary grain $A' \sim X_h^+(f')$.

In addition, in the case of the Boolean random function, one has

$$D_Z(\tau) = \bigcup_{x_i} [D_{f_i}(\tau)]_{x_i} \Rightarrow [Z \triangledown^* \check{\tau}^c]_+ = \left[\bigvee_{x_i} f_i(x - x_i) \triangledown^* \check{\tau}^c \right]_+,$$

such that $D_Z(\tau)$ is a Boolean RACS with the primary grain $D_{f'}(\tau) = [f' \triangledown^* \check{\tau}^c]_+$. This property is very useful in practice for the identification of a model of Boolean random function using tools for Boolean RACS models. The expression of the Choquet capacity for a Boolean random function has also a nice expression.

Proposition 23. *Given a Boolean random function of parameters (θ, f'), its Choquet capacity is given by (Jeulin, 1991)*

$$1 - T_Z(\tau) = Q_Z(\tau) = \exp\left(-\theta \mathbb{E}\left\{\mu_n\left([f' \triangledown^* \check{\tau}^c]_+\right)\right\}\right),$$

and therefore, the logarithm of the characteristic functional is given by

$$\log Q_Z(\tau) = -\theta \mathbb{E}\left\{\mu_n\left([f' \triangledown^* \check{\tau}^c]_+\right)\right\}. \tag{119}$$

Proof. This proof is inspired from the one of the Boolean model provided in Lantuéjoul (1993). Let us denote by $X(B)$ the union of the primary grains implanted in a bounded domain B. The number of primary grains follows a Poisson distribution of parameter $\theta\mu_d(B)$. Take a uniform point $x \in B$ and implant a primary function at x, denoted f'_x. It is "disjoint" with τ with probability

$$P\{x \notin D_{f'}(\tau)\} = 1 - \frac{\mathbb{E}\{\mu_d(D_{f'}(\tau) \cap B)\}}{\mu_d(B)}.$$

When the number of points equals i, the i primary functions has been independently and uniformly implanted. Consequently,

$$Q_Z^B(\tau) = \sum_{i=0}^{+\infty} e^{-\theta\mu_d(B)} \frac{(\theta\mu_d(B))^i}{i!} \left(1 - \frac{\mathbb{E}\{\mu_d(D_{f'}(\tau) \cap B)\}}{\mu_d(B)}\right)^i.$$

This sum is equal to

$$Q_Z^B(\tau) = e^{-\theta \mathbb{E}\left\{\mu_d\left(D_{f'}(\tau)\cap B\right)\right\}}.$$

The results should be extended to complete domain of definition of Z. Let us consider an increasing sequence B_n of bounded domains in $\mathbb{R}^d$. The events "$K \cap X(B_n) = \emptyset$" of probability $Q_Z^n(K)$ are an increasing sequence of intersection "$K \cap X = \emptyset$". By the axiom of continuity we have

$$Q_Z(K) \quad = \quad \lim_{n\to+\infty} Q_Z^n(K) = \lim_{n\to+\infty} e^{-\theta E\left\{\mu_d\left(D_{f'}(\tau)\cap B_n\right)\right\}} = e^{-\theta \mathbb{E}\left\{\mu_d\left(D_{f'}(\tau)\right)\right\}}.$$

Finally, from (117), one has

$$D_{f'}(\tau) = \left[f' \,\nabla^* \,\check{\tau}^c\right]_+.$$

□

From the expression (119), different particular cases τ can be used to obtain for instance the spatial law of the Boolean random function, its bivariate distribution, etc. (Jeulin, 1991).

A particular case corresponds to the case of a test function of type inverted cylinder of height $k \in \mathbb{R}$ and base K:

$$\tau(x) = \{K; k\} = \begin{cases} k & \text{if } x \in K \\ +\infty & \text{if } x \notin K \end{cases}$$

such that one has

$$[D_Z(\tau)]^c = \left\{x : \bigvee_{y\in K} Z(x+y) < k\right\}.$$

This expression can be rewritten in morphological terms as

$$\bigvee_{y\in K} Z(y) < k \Leftrightarrow K_x \subset Y_k^-(Z) \Leftrightarrow x \in \left(Y_k^-(Z) \ominus K\right),$$

and the complementary event

$$\bigvee_{y\in K} Z(y) \geq k \Leftrightarrow x \in \left[Y_k^-(Z) \ominus K\right]^c = X_k^+(Z) \oplus \check{K} = D_Z(\tau).$$

For this type of test function $\{K; k\}$, we have in the Boolean case

$$P\left\{\bigvee_{x\in K} Z(x) < k\right\} = \exp\left(-\theta\mathbb{E}\left\{\mu_n\left(X_k^+(f') \oplus \check{K}\right)\right\}\right).$$

It gives the probability distribution of the Boolean random function $Z(x)$ after a change of support by supremum over the compact set K.

We believe that the interest of the expression (119) of $Q_Z(\tau)$, based on a (max, min)-convolution, opens avenues regarding a more widely use of Boolean random functions. In particular, we can see how the role of convexity of the primary grain A' in the Boolean model is replaced here by the convexity of the upper level sets of the primary function f', which involves that the most interesting case corresponds to quasiconcave primary functions f'. Other issues, as the interest of the complement of the conic structuring function $c^c_{\alpha,\lambda}(x) = \lambda^{-1}\|x\| + \alpha$ as privileged test function, as well as the simulation of Boolean random function using the Hamilton–Jacobi PDEs discussed above, will be considered in ongoing research.

Bibliographic remark. In the initial theory of Boolean random functions, it was classically considered as involving the Minkowski addition of the subgraph of function $Z(x)$ by compact test sets K, where K is seen as a subset of $\mathbb{R}^n \times \mathbb{R}$ (Serra, 1989; Préteux & Schmitt, 1988), which is not equivalent to the use of structuring functions. The formulation based on test functions and level-set representation of the Choquet capacity introduced by Jeulin (1991) has allowed us to link Boolean random function characterization with adjoint supmin convolution.

We believe that most results proved by Goutsias (1992) for the discrete framework of RACS can be now extended to discrete random function.

Let us note that if we consider the case of Boolean functions defined in the unit interval [0, 1], the immediate consequence is the fact that the underlaying framework can be seen as the appropriate one to model Boolean fuzzy sets. Remember that the key ingredient is the implication of the Gödel–Brower Logic. Up to the best of our knowledge, no previous work has ever studied the interest of fuzzy morphological operators to model Boolean fuzzy sets. This point seems now natural and deserves a deeper effort of research.

Concerning just the theory of random fuzzy closed sets, some efforts have been paid recently in order to generalize Matheron's results for RACS to fuzzy sets. In particular, there are already some relevant results including

Choquet theorem for random USC functions based on Lawson topology in the continuous lattice theory, that includes also some tentative of generalization of Matheron's hit-or-miss topology. However many topological questions, including measurability considerations, are still open (Nguyen, Wang, & Wei, 2007; Wei & Wang, 2007). In this context, we think that the Lawson topology of continuous lattices (Gierz et al., 1980) is also relevant for the study of the topological properties of (max, min)-convolutions.

6.6 Links with Geodesic Dilation and Erosion

In Section 2, on the background on classical morphological operators, we did not consider the notion of geodesic dilation and erosion (Lantuéjoul & Beucher, 1981; Vincent, 1993). Geodesic operators are extremely useful transforms for solving many image processing tasks (Vincent, 1993; Soille, 1999).

The philosophy of these operators is rather different of the classical convolution-like dilation and erosion, since the role of the structuring element/function is replaced by an external function called the reference (or geodesic mask). More precisely, the elementary geodesic dilation $\delta_X^{1-geo}(Y)$ of a set Y included in the geodesic mask X, $X, Y \subset \mathbb{R}^n$, is defined as

$$\delta_X^{1-geo}(Y) = (Y \oplus B) \cap X, \tag{120}$$

where B is the unit Euclidean ball. The elementary geodesic erosion $\varepsilon_X^{1-geo}(Y)$ is defined by set complementation:

$$\varepsilon_X^{1-geo}(Y) = X \setminus \delta_X^{1-geo}(X \setminus Y) = ((Y \cup X^c) \ominus B) \cap X. \tag{121}$$

We note that geodesic dilation and erosion for sets are dual both by complement and by adjunction. Larger operations are obtained by iteration and keeping X fixed. Thus, the geodesic dilation of size n of Y corresponds to the space swept by the geodesic balls $B_X(x, n)$ of size n whose center y is included in Y; the geodesic erosion of size n corresponds to the centers of the geodesic balls of size n of X included in Y (Lantuéjoul & Beucher, 1981), i.e.,

$$\begin{aligned} \delta_X^{n-geo}(Y) &= \{x \in X : B_X(x, n) \cap Y \neq \emptyset\} = \bigcup_{y \in Y} B_X(y, n), \\ \varepsilon_X^{n-geo}(Y) &= \{x \in X : B_X(x, n) \subset Y\}. \end{aligned}$$

Extension from the set operators to the geodesic operators for functions is based on a level set based transposition of set geodesic dilation, which leads to the geodesic dilation $\delta_g^{1-geo}(f)$ of a function f in the geodesic space of g (or under g, since it is assumed that $f \le g$):

$$\delta_g^{1-geo}(f) = \left(f \oplus B\right) \wedge g,\ f \le g. \tag{122}$$

The geometric interpretation is similar to the one for the set case, by considering the notion of sweeping geodesic cylinders. Then, the geodesic erosion is classically defined using the duality by complement from (122) to obtain (Vincent, 1993):

$$\varepsilon_g^{1-geo}(f) = \left(f \ominus B\right) \vee g,\ f \ge g. \tag{123}$$

In fact, it is easy to see that the functional geodesic erosion (123) is not the operator obtained by the level set extension of the set geodesic erosion (121). That involves in fact that $\varepsilon_g^{1-geo}(f)$ and $\delta_g^{1-geo}(f)$ do not form an adjunction. It is very surprising for us that the study of the geodesic adjunction has been the object of depth research only in a relatively recent and illuminating study by Beucher (2011).

Starting from the dual by complement operators (122) and (123), computation of the corresponding adjoint operators is provided in detail in Beucher (2011). The four elementary geodesic operators are given by

$$\begin{array}{lclcl}
g \le f, & \overline{\varepsilon}_g^{1-geo}(f) = \left(f \ominus B\right) \vee g & \overset{\text{adjoint}}{\longleftrightarrow} & \overline{\delta}_g^{1-geo}(f) = \left[\left(f \wedge m\right) \oplus B\right] \vee g \\
 & & & \text{with } m = \{x : f > g\} \\
 & \updownarrow \text{dual} & & \updownarrow \text{dual} \\
g \ge f, & \underline{\delta}_g^{1-geo}(f) = \left(f \oplus B\right) \wedge g & \overset{\text{adjoint}}{\longleftrightarrow} & \underline{\varepsilon}_g^{1-geo}(f) = \left[\left(f \vee m\right) \ominus B\right] \wedge g \\
 & & & \text{with } m = \{x : f = g\}.
\end{array}$$

In order to relate geodesic operators to (max, min) algebra, we can rewrite the geodesic dilation $\underline{\delta}_g^{1-geo}(f)$ using the distributivity:

$$\begin{aligned}
\underline{\delta}_g^{1-geo}(f)(x) &= \left(f \oplus B\right)(x) \wedge g(x) = g(x) \wedge \sup_{z \in B(x)} f(x-z) \\
&= \sup_{z \in B(x)} \left\{g(x) \wedge f(x-z)\right\}.
\end{aligned} \tag{124}$$

We can also find an alternative expression for its adjoint geodesic erosion $\underline{\delta}_g^{1-geo}(f)$, with $f \le g$:

$$\underline{\varepsilon}_g^{1-geo}(f)(x) = \left[\left(f \vee m\right) \ominus B\right](x) \wedge g(x)$$

$$= \inf_{z \in B(x)} \left\{ g(x) \wedge \left(f(x-z) \vee m(x-z) \right) \right\}.$$

First, we note that using the mask function given by $m(x) = \top$ if $f(x) = g(x)$ and $m(x) = \perp$ if $f(x) < g(x)$, one has $(f \vee m)(x) = \top$ if $f(x) = g(x)$ and $(f \vee m)(x) = f(x)$ if $f(x) < g(x)$. Second, we introduce the operator $\wedge^{*-geo}$ as

$$g(x) \wedge^{*-geo} f(x-z) = g(x) \wedge \left(f(x-z) \vee m(x-z) \right)$$
$$= \begin{cases} g(x) & \text{if } f(x-z) = g(x-z) \\ f(x-z) \wedge g(x) & \text{if } f(x-z) < g(x-z) \end{cases}$$

Thus, we finally obtain

$$\underline{\varepsilon}_g^{1-geo}(f)(x) = \inf_{z \in B(x)} \left\{ g(x) \wedge^{*-geo} f(x-z) \right\}. \tag{125}$$

The operator $\wedge^{*-geo}$ can be seen as the "geodesic counterpart" of the adjoint operator to the minimum (24). Going further with the parallelism, we believe that, even if the geodesic dilation and its adjoint geodesic erosion are not convolution-like operators, it is seems clear from expressions (124) and (125) that these two operators are structurally similar respectively to the supmin convolution (20) and the adjoint infmax convolution (22). An equivalent interpretation is straightforward obtained for the geodesic erosion $\overline{\varepsilon}_g^{1-geo}(f)$ and its adjoint geodesic dilation $\overline{\delta}_g^{1-geo}(f)$.

Thus, we would like to conclude by stating that geodesic dilation and erosion are naturally defined as operators in (max, min)-algebra. We plan to fully exploit this new viewpoint in future research, including links of geodesic operators with nonlinear analysis and Hamilton–Jacobi PDEs.

7. CONCLUSION AND PERSPECTIVES

Operators and filters underlying a formulation as (max, min)-convolutions are common in the state-of-the-art of mathematical morphology. However, their study *per se* has been neglected. From this epistemological viewpoint, we can conclude that the role of (max, min)-convolutions has been somewhat overshadowed by a multiplicity of viewpoints (fuzzy, viscous, "hitting of functions" in Choquet capacity, etc.).

In order to address this theoretical lack, we have developed in our paper a rigorous formulation and characterization of the four convolution-like operators in (max, min)-algebra.

Just concerning the need of specifically four morphological operators in such algebra, we point out that this is clearly justified by the fact that one deals with two dualities, by complement and by adjunction. This principle is not exclusive from mathematical morphology, in optimization theory involving minimum or maximum operations, it is also a natural principle well formalized (Flachs & Pollatschek, 1979).

We have proved that (max, min)-openings are compatible with Matheron's axiomatic of Euclidean granulometries for functions with quasiconcave structuring functions. This was not the case for (max, +)-openings. We have also shown that the adjoint supmin convolution is the operator underlying the extension of Matheron's characterization of Boolean random closed sets to the case of Boolean random upper semicontinuous function.

For all these reasons, we strongly think that (max, min)-convolution provides the natural framework to generalize some key notions of Matheron's theory from sets to functions.

All the theoretical results provided in the paper are formulated in a continuous setting. Although it may seem that the transition from the continuous to the discrete case is straightforward (this is usually the case in the theory of mathematical morphology), this transition is often challenging with a certain number of notions, such as convexity. We plan to revisit all our results in a fully discrete framework.

All the algebraic results (and most of nonlinear analysis ones) on (max, min)-convolutions considered here are valid for functions supported in a general Banach space, consequently more general that the Euclidean space $\mathbb{R}^n$. In this generalization context, we plan to consider in particular the case of (max, min)-morphology for real-valued images on Riemannian manifolds. It will be a parallel study the our recent work on the extension of (max, +)-morphology to Riemannian images (Angulo & Velasco-Forero, 2014).

A final remark concerning the generalization of the morphological convolutions proposed here. If we compare for instance the classical erosion and the infmax erosion:

$$
\begin{aligned}
(f \ominus b)(x) &= \inf_y \left\{ f(y) + \check{b}^c(x-y) \right\}, \\
(f \bigtriangleup b)(x) &= \inf_y \left\{ f(y) \vee \check{b}^c(x-y) \right\},
\end{aligned}
$$

they can be considered as a particular case of the following transformation:

$$(f \overline{\wedge}_p b)(x) = \inf_y \left\{ N_p \left(f(y), \check{b}^c(x-y) \right) \right\}, \tag{126}$$

where $N_p(r, s)$ is a monotone norm, for instance the L^p Minkowski norm; i.e., for $p \geq 1$

$$N_p = (|r|^p + |s|^p)^{\frac{1}{p}},$$

such that $(f \ominus b) = (f \overline{\wedge}_1 b)$ and $(f \vartriangle b) = (f \overline{\wedge}_\infty b)$. Besides these limit cases, the study of the properties of the operator (126) and its dual and adjoint counterparts is potentially relevant in mathematical morphology theory. Some results already obtained in the domain of convex analysis, see Volle (1994), can be a good starting point.

In previous work, we have considered a nonlinearization of continuous diffusion (Angulo, 2016) and bilateral discrete filtering (Angulo, 2013) in order to produce families of processed images from (Gaussian) convolution in $(+, \times)$-algebra to (parabolic) morphological operators in $(\max, +)$-algebra. The development of a computational framework which unifies the three algebras, from (Gaussian) convolution in $(+, \times)$-algebra to (conic) morphological operators in $(\max, \min)$-algebra would be very powerful for the efficient implementation of generic convolution in specific hardware/software architectures.

Concerning the Hamilton–Jacobi PDEs associated to supmin and infmax convolutions, we have not discussed their numerical solution. Numerical algorithms for Hamilton–Jacobi PDE are relatively well known, in particular the Osher and Sethian algorithm (Osher & Sethian, 1988) or the less diffusive Rouy and Tourin schema (Rouy & Tourin, 1992). Alternative schemas can be also found in Maragos (2003), Breuß and Weickert (2009), Diop and Angulo (2015). Nevertheless, ongoing research on the behavior of such numerical approaches is needed to state the accuracy, convergence, etc. for different structuring functions, i.e., different Hamiltonians. As usual, the interest of PDE approaches is to provide infinitesimal computation with a geometrically better approximation of the Euclidean distance. In addition, in the case of $(\max, \min)$-convolutions framework, it is easy to implement many different filtering effects only by changing the Hamiltonian. On the contrary, we should notice that the PDE approach is not appropriate for implementing advanced operators; i.e., alternate sequential filters obtained by the composition of openings and closings at various scales.

We sincerely hope that results reviewed in Sections 4 and 5 will help a general reader to appreciate the fact that applied mathematics underlying morphological operators are well founded and pertinent from nonlinear analysis. Theoretical roots of mathematical morphology are nowadays formulated in the framework of algebra (more specifically, using complete lattice theory): we expect that the results provided in the paper justify a complementary coherent theoretical viewpoint of morphological operators from nonlinear analysis.

ACKNOWLEDGMENT

I would like to thank F. Meyer for fruitful discussions on this work. Many thanks also to D. Jeulin for his precisions on Boolean random function characterization.

REFERENCES

Alvarez, O., Barron, E. N., & Ishii, H. (1999). Hopf–Lax formulas for semicontinuous data. *Indiana University Mathematics Journal, 48*, 993–1035.

Alvarez, L., Guichard, F., Lions, P.-L., & Morel, J.-M. (1993). Axioms and fundamental equations of image processing. *Archive for Rational Mechanics and Analysis, 123*(3), 199–257.

Angulo, J. (2013). Morphological bilateral filtering. *SIAM Journal on Imaging Sciences, 6*(3), 1790–1822.

Angulo, J. (2016). Generalised morphological image diffusion. *Nonlinear Analysis: Theory, Methods & Applications, 134*, 1–30.

Angulo, J., & Velasco-Forero, S. (2013). Stochastic morphological filtering and Bellman–Maslov chains. In *Lecture notes in computer science: Vol. 7883. Proceedings of ISMM'13 (11th international symposium on mathematical morphology)* (pp. 171–182). Springer-Verlag.

Angulo, J., & Velasco-Forero, S. (2014). Riemannian mathematical morphology. *Pattern Recognition Letters, 47*, 93–101.

Arehart, A. B., Vincent, L., & Kimia, B. B. (1993). Mathematical morphology: The Hamilton–Jacobi connection. In *Proceedings of IEEE 4th international conference on computer vision (ICCV'93)* (pp. 215–219).

Attouch, D., & Aze, D. (1993). Approximation and regularization of arbitrary functions in Hilbert spaces by the Lasry–Lions method. *Annales de l'Institut Henri Poincaré, Section C, 10*(3), 289–312.

Avriel, M., Diewert, W. E., Schaible, S., & Zang, I. (1988). *Generalized concavity*. Plenum Press.

Bardi, M., & Evans, L. C. (1984). On Hopf's formulas for solutions of Hamilton–Jacobi equations. *Nonlinear Analysis: Theory, Methods & Applications, 8*(11), 1373–1381.

Barron, E. N., Jensen, R., & Liu, W. (1996). Hopf–Lax formula for $u_t = H(u, Du) = 0$. *Journal of Differential Equations, 126*, 48–61.

Barron, E. N., Jensen, R., & Liu, W. (1997). Hopf–Lax formula for $u_t = H(u, Du) = 0$, II. *Communications in Partial Differential Equations, 22*, 1141–1160.

Beucher, S. (2011). *About a problem of definition of the geodesic erosion* (CMM/Mines ParisTech technical report). Retrieved from http://cmm.ensmp.fr/~beucher/publi/GeodesicErode_eng.pdf.

Bloch, I. (2009). Duality vs. adjunction for fuzzy mathematical morphology and general form of fuzzy erosions and dilations. *Fuzzy Sets and Systems, 160*, 1858–1867.

Breuß, M., & Weickert, J. (2009). Highly accurate PDE-based morphology for general structuring elements. In *Lecture notes in computer science: Vol. 5567. Proceedings of scale space and variational methods in computer vision* (pp. 758–769). Berlin: Springer.

Brockett, R. W., & Maragos, P. (1994). Evolution equations for continuous-scale morphology. *IEEE Transactions on Signal Processing, 42*(12), 3377–3386.

Clarke, F. H., & Stern, R. J. (Eds.). (1999). *NATO science series C: Mathematical and physical sciences: Vol. 528. Nonlinear analysis, differential equations and control*. Kluwer.

Crandall, M. G., Ishii, H., & Lions, P.-L. (1992). User's guide to viscosity solutions of second order partial differential equations. *Bulletin of the American Mathematical Society, 27*(1), 1–67.

Darbon, J. (2015). On convex finite-dimensional variational methods in imaging sciences and Hamilton–Jacobi equations. *SIAM Journal on Imaging Sciences, 8*(4), 2268–2293.

Deng, T.-Q., & Heijmans, H. J. A. M. (2002). Grey-scale morphology based on fuzzy logic. *Journal of Mathematical Imaging and Vision, 16*(2), 155–171.

Diop, E. H. S., & Angulo, J. (2015). Multiscale image analysis based on robust and adaptive morphological scale-spaces. *Image Analysis & Stereology, 34*(1), 39–50.

Dorst, L., & van den Boomgaard, R. (1994). Morphological signal processing and the slope transform. *Signal Processing, 38*, 79–98.

Engbers, E. A., Van Den Boomgaard, R., & Smeulders, A. W. M. (2001). Decomposition of separable concave structuring functions. *Journal of Mathematical Imaging and Vision, 15*(3), 181–195.

Flachs, J., & Pollatschek, M. A. (1979). Duality theorems for certain programs involving minimum or maximum operations. *Mathematical Programming, 16*, 348–370.

Gierz, G., Hofmann, K. H., Keimel, K., Lawson, J. D., Mislove, M., & Scott, D. S. (1980). *A compendium of continuous lattices*. Berlin: Springer-Verlag.

Gondran, M. (1996). Analyse MINMAX. *Comptes Rendus de l'Académie des Sciences, Série I, 323*, 1249–1252.

Gondran, M., & Minoux, M. (2008). *Graphs, dioids and semirings: New models and algorithms*. Springer.

Goutsias, J. (1992). Morphological analysis of discrete random shapes. *Journal of Mathematical Imaging and Vision, 2*, 193–215.

Heijmans, H. J. A. M. (1994). *Morphological image operators*. Boston: Academic Press.

Jackway, P. T., & Deriche, M. (1996). Scale-space properties of the multiscale morphological dilation-erosion. *IEEE Transactions on Pattern Analysis and Machine Intelligence, 18*(1), 38–51.

Jeulin, D. (1991). *Modèles morphologiques de structures aléatoires et de changement d'echelle* (Thèse d'Etat dès sciences physiques). France: Université de Caen.

Jeulin, D. (2014). Boolean random functions. In V. Schmidt (Ed.), *Lecture notes in mathematics: Vol. 2120. Stochastic geometry, spatial statistics and random fields*. Springer (Chapter 5).

Jeulin, D., & Jeulin, P. (1981). Synthesis of rough surfaces by random morphological model. Proceedings of the 3rd European symposium of stereology. *Stereologia Iugoslavica, 3*(Suppl. 1), 239–246.

Kiselman, C. (2007). Division of mappings between complete lattices. In *Proceedings of the 8th international symposium on mathematical morphology (ISMM'07), Vol. 1* (pp. 27–38). MCT/INPE.

Kiselman, C. (2010). Inverses and quotients of mappings between ordered sets. *Image and Vision Computing, 28*, 1429–1442.

Kolokoltsov, V., & Maslov, V. P. (1997). *Idempotent analysis and its applications*. Springer.

Kraus, E. J., Heijmans, H. J. A. M., & Dougherty, E. R. (1993). Gray-scale granulometries compatible with spatial scalings. *Signal Processing, 34*(1), 1–17.

Lantuéjoul, Ch. (1993). *Cours "Les ensembles aléatoires"* (Lecture notes). Ecole des Mines de Paris. Retrieved from cg.ensmp.fr/bibliotheque/public/LANTUEJOUL_Cours_00578.pdf.

Lantuéjoul, Ch., & Beucher, S. (1981). On the use of the geodesic metric in image analysis. *Journal of Microscopy, 121*(Part 1), 39–49.

Lasry, J. M., & Lions, P.-L. (1986). A remark on regularization in Hilbert spaces. *Israel Journal of Mathematics, 55*, 257–266.

Luc, D. T., & Volle, M. (1997). Levels sets infimal convolution and level addition. *Journal of Optimization Theory and Applications, 94*(3), 695–714.

Maragos, P. (1995). Slope transforms: Theory and application to nonlinear signal processing. *IEEE Transactions on Signal Processing, 43*(4), 864–877.

Maragos, P. (1996). Differential morphology and image processing. *IEEE Transactions on Image Processing, 5*(1), 922–937.

Maragos, P. (2003). Algebraic and PDE approaches for lattice scale-spaces with global constraints. *International Journal of Computer Vision, 52*(2/3), 121–137.

Maragos, P. (2005). Lattice image processing: A unification of morphological and fuzzy algebraic systems. *Journal of Mathematical Imaging and Vision, 22*(2–3), 333–353.

Maragos, P. (2013). *Representations for morphological image operators and analogies with linear operators. Advances in imaging and electron physics: Vol. 177*. Elsevier.

Maragos, P., & Schafer, R. W. (1987). Morphological filters. Part I: Their set-theoretic analysis and relations to linear shift-invariant filters. *IEEE Transactions on Acoustics, Speech, and Signal Processing, 35*(8), 1153–1169.

Maragos, P., & Vachier, C. (2008). A PDE formulation for viscous morphological operators with extensions to intensity-adaptive operators. In *Proceedings of the 15th IEEE international conference on image processing (ICIP'08)* (pp. 2200–2203).

Matheron, G. (1975). *Random sets and integral geometry*. New York: Wiley.

Meyer, F. (1993). *Inondation par des fluides visqueux* (CMM/Mines ParisTech technical report).

Meyer, F. (2008). *Operateurs morphologiques visqueux* (CMM/Mines ParisTech technical report).

Moreau, J. J. (1970). Inf-convolution, convexité des fonctions numériques. *Journal de Mathématiques Pures et Appliquées, 49*, 109–154.

Nachtegael, M., & Kerre, E. E. (2001). Connections between binary, gray-scale and fuzzy mathematical morphologies. *Fuzzy Sets and Systems, 124*, 73–85.

Nguyen, H. T., Wang, Y., & Wei, G. (2007). On Choquet theorem for random upper semicontinuous functions. *International Journal of Approximate Reasoning, 46*(1), 3–16.

Osher, S., & Sethian, J. (1988). Fronts propagating with curvature-dependent speed: Algorithms based on Hamilton–Jacobi formulations. *Journal of Computational Physics, 79*, 12–49.

Penot, J.-P., & Zălinescu, C. (2001). Approximation of functions and sets. In M. Lassonde (Ed.), *Approximation, optimization and mathematical economics* (pp. 255–274). Heidelberg: Physica-Verlag.

Préteux, F., & Schmitt, M. (1988). Boolean texture analysis and synthesis. In J. Serra (Ed.), *Image analysis and mathematical morphology, Vol. 2*. Academic Press (Chapter 18).

Ritter, G., & Wilson, J. (2011). *Handbook of computer vision algorithms and image algebra*. CRC.

Rockafellar, R. T. (1970). *Convex analysis*. Princeton, NJ: Princeton University Press.

Rouy, E., & Tourin, A. (1992). A viscosity solutions approach to shape from shading. *SIAM Journal on Numerical Analysis, 29*(3), 867–884.

Schmitt, M. (1996). Support function and Minkowski addition of non-convex sets. In P. Maragos, R. Shafer, & M. Butt (Eds.), *Mathematical morphology and its applications to image and signal processing* (pp. 15–22). Kluwer.

Seeger, A., & Volle, M. (1995). On a convolution operation obtained by adding level sets: Classical and new results. *Operations Research, 29*(2), 131–154.

Serra, J. (1982). *Image analysis and mathematical morphology*. London: Academic Press.

Serra, J. (1988). *Image analysis and mathematical morphology. Vol. II: Theoretical advances*. London: Academic Press.

Serra, J. (1989). Boolean random functions. *Journal of Microscopy, 156*(Pt. 1), 41–63.

Soille, P. (1999). *Morphological image analysis*. Berlin: Springer-Verlag.

Vachier, C., & Meyer, F. (2005). The viscous watershed transform. *Journal of Mathematical Imaging and Vision, 22*, 251–267.

Vachier, C., & Meyer, F. (2007). News from viscous land. In *Proceedings of the 8th international symposium on mathematical morphology (ISMM'07)* (pp. 189–200).

Van, T. D., & Son, N. D. T. (2006). Hopf–Lax–Oleinik-type estimates for viscosity solutions to Hamilton–Jacobi equations with concave–convex data. *Vietnam Journal of Mathematics, 34*(2), 209–239.

van den Boomgaard, R., & Dorst, L. (1997). The morphological equivalent of Gaussian scale-space. In *Proceedings of Gaussian scale-space theory* (pp. 203–220). Kluwer.

Vincent, L. (1993). Morphological grayscale reconstruction in image analysis: Applications and efficient algorithms. *IEEE Transactions on Image Processing, 2*(2), 176–201.

Volle, M. (1994). The use of monotone norms in epigraphical analysis. *Journal of Convex Analysis, 1*(2), 203–224.

Volle, M. (1998). Duality for the level sum of quasiconvex functions and applications. *ESAIM. Control, Optimisation and Calculus of Variations, 3*, 329–343.

Wei, G., & Wang, Y. (2007). On metrization of the hit-or-miss topology using Alexandroff compactification. *International Journal of Approximate Reasoning, 46*(1), 47–64.

CHAPTER TWO

Critical Magnetic Field and Its Slope, Specific Heat, and Gap for Superconductivity as Modified by Nanoscopic Disorder

Clifford M. Krowne
Naval Research Laboratory, Washington, DC, United States
e-mail address: cliff.krowne@nrl.navy.mil

Contents

1. INTRODUCTION

Superconductivity can be modified by various effects related to randomness, disorder, structural defects, and other similar physical effects. Their affects on superconductivity are important because such effects are intrinsic to certain material system's preparation, or may be intentionally produced. Here we show that the introduction of disorder through impurities could possibly lead to an increase in the superconducting gap $\Delta(T)$, where this disorder is of an unconventional non-alloy type. Furthermore,

Advances in Imaging and Electron Physics, Volume 203
ISSN 1076-5670
http://dx.doi.org/10.1016/bs.aiep.2017.07.004

by use of this $\Delta(T)$ in a microscopic approach, critical magnetic field H_c, its slope dH_c/dT, and heat capacity C_{el} are found.

Superconductivity may be able to play a new or improved role in advancing technologies such as wireless communications, satellites, and other electronic and electromagnetic technologies which require the use of microscopic or nanoscopic materials, structures, and devices. These all enlist small superconductors in 3D, 2D, 1D, and 0D, which includes bulk like systems, layered or thin film or atomic sheet systems (Osofsky, Hernández-Hangarter, et al., 2016; Osofsky, Krowne, et al., 2016), nanowires/nanotubes/nanocables (Krowne, 2011; Martinez, Calle-Vallejo, Krowne, & Alonso, 2012; Martinez, Abad, Calle-Vallejo, Krowne, & Alonso, 2013; Abrahams, Anderson, Licciardello, & Ramakrishnan, 1979), and quantum dots. Obtaining superconductors in such small systems necessitates relaxed requirements on critical values of temperature T_c, current J_c, and magnetic field H_{c1} and H_{c2}. One way to obtain larger values of all these critical constants, may be to introduce disorder intentionally, of an unconventional non-alloy type.

We have found that functionalizing graphene (Osofsky, Hernández-Hangarter, et al., 2016) with atoms of F, N, or O, allows one to dial into various matter states, which may be metallic, insulating, or even superconducting. This ability to dial in such characteristics at the nanoscopic or atomic level is something that was not attainable previously in a systematic fashion, which can be repeatedly duplicated now in the laboratory. This is the latest era of atomic and nanoscopic science: dimensions manipulated are on the order of a few angstroms. One of the remarkable facets of that study is despite the excitement generated by the achievement of metallic single layer graphene, it has been confused by the fact that seminal theoretical work (Abrahams et al., 1979) predicted that purely two-dimensional (2D) systems should not be metallic – because of disorder. The situation is confounded further by later theoretical work showing that Dirac Fermionic systems with no spin–orbit interactions and Gaussian correlated disorder exhibit scaling behavior, should always be metallic (Das Sarma, Adam, Hwanf, & Rossi, 2011). So what is it, metallic or insulating? Really the answer is both due to the metal–insulator transition (MIT), which also suggests the possibility of dialing into a superconducting state.

Extremely thin metallic-like oxides, on the scale of ten nanometers, such as RuO_2 thin film layers (Osofsky, Krowne, et al., 2016), is a highly disordered conductor in which resistivity does not decrease with decreasing temperature, and the disorder-driven MIT is then described as a quantum phase transition characterized by extended states for the metallic phase and by localized states for the insulating phase. The existence of metallic behavior reported for this 2D material once again violates the famous prediction of Abrahams et al. (1979) that all 2D systems must be localized regardless of the degree of disorder. The discovery of a metallic state in high-mobility metal oxide field-effect transistors (HMFET) motivated several theoretical approaches that included electron–electron interactions to screen disorder. These models adequately described the results for low carrier concentration, high-mobility systems, but are not applicable to the case of highly disordered 2D metals. And once again, the metal–insulator transition (MIT), suggests the possibility of dialing into a superconducting state.

Dimensions for RuO_2 nanoparticles which decorate SiO_2 inner core nanowires, forming nanocables (Martinez et al., 2013), are on the order of 3 nm. These 1D type of structures, might be investigated for possibilities of superconductivity. Although modeling makes some substantial simplifications, looking at the continuous cylindrical coverage of RuO_2 as a nanotube, in fact it is composed of somewhat randomly placed popcorns of RuO_2 nodules. Although early papers, for example, the one by Anderson (1959), have made some qualitative arguments for disorder effects based upon some first and second order quantum mechanical perturbation theory, the theoretical and experimental insights dependent upon atomic and nanoscopic engineering of materials, did not exist back then, and so limits the reach of such earlier works.

However, ultrasmall samples used in tunneling experiments, led to studies calling into question use of the grand canonical ensemble to describe electron pairing gap $\Delta_0(n_e)$, using an attractive Hubbard model (in real space) to represent s-wave superconductors, obtaining the gap parameter $\Delta_{N_e}(n_e, N)$ varying with averaged electron density n_e and site number N (Tanaka & Marsiglio, 2000a). In Tanaka and Marsiglio (2000a), and in Tanaka and Marsiglio (2000b), the suggested approach in Anderson (1959), solving for the eigenvalues and eigenstates of a non-interacting problem, diagonalizing the single-particle Hamiltonian, finding the transformed electron–electron interaction, and then applying the BCS variational procedure, generates a modified BCS gap equation. An alternative view using the effective Hamiltonian, diagonalizing by a Bogoliubov–Valatin transformation with a de Gennes approach, allows inspection of specific sites with one added impurity atom (Tanaka & Marsiglio, 2000b). Such studies point up the importance of looking at the various nanoscopic aspects of superconductivity, although in the work here, we will not address such site by site nonuniformities (with either BCS and/or BdG tactic).

Superconductivity can be modified by various effects related to randomness, disorder, structural defects, and other similar physical effects. Their affects on superconductivity is important because such effects are intrinsic to certain material system's preparation, or may be intentionally produced. For example, in the 1980s on Anderson localization and dirty superconductors H_{c2} is affected (Kotliar & Kapitulnik, 1986); and high T_c superconductors with structural disorder affects the electron–electron attraction stipulated by exchange of low-energy excitations, with substantial enhancement of T_c (Maleyev & Toperverg, 1988). In the 1990s, looking at thermal fluctuations, quenched disorder, phase transitions, and transport in type-II superconductors, vortices are pinned due to impurities or other defects which destroys long range correlations of the vortex lattice (Fisher, Fisher, & Huse, 1991); in field-induced superconductivity in disordered wire networks, small transverse magnetic applied fields increased the mean T_c in disordered networks (Bonetto, Israeloff, Pokrovskiy, & Bojko, 1998); for structural disorder and its effect on the superconducting transition temperature in the organic superconductor κ-(BEDT-TTF)$_2$Cu[N(CN)$_2$]Br, T_c is reduced in quenched cooled state (Su, Zuo, Schlueter, Kelly, & Williams, 1998); enhancement of J_c density in single-crystal $Bi_2Sr_2CaCu_2O_8$ superconductors occurs by chemically induced disorder (Wang, Wu, Chen, & Lieber, 1990). In the 2000s, surface enhancement of superconductivity occurred in single crystal tin, due to cold worked surface with surface enhanced order parameter (Kozhevnikov et al., 2005); for disorder and quantum fluctuations in superconducting films in strong magnetic fields, H_{c2} can increase and especially at low temperature (Galitski & Larkin, 2001); preparing amorphous MgB_2/MgO superstructures which produces a model disordered superconductor, bilayers were made with relatively high T_c (Siemons et al., 2008); disorder-induced superconductivity can be produced in ropes of carbon nanotubes, with T_c increasing with disorder (Bellafi, Haddad, & Charfi-Kaddour, 2009); disordered 2D superconductors are examined for the role of temperature and interaction strength in the Hubbard model when the on-site attraction is switched off on a fraction f of sites while keeping the attraction U on the remaining sites, showing that near $f = 0.07$, T_c increases with U in the $2 \leq U \leq 6$ range (Mondaini, Paiva, dos Santos, & Scalettar, 2008); enhancement of the high magnetic field J_c of superconducting MgB_2 by proton irradiation occurs, with the irradiation pinning the vortices increasing J_c (Bugoslavsky et al., 2001); Examination of the Lindemann criterion and vortex phase transitions in type-II superconductors, shows the destruction of vortex order by random point pinning and thermal fluctuations (Kierfeld & Vinokur, 2004); doping induced disorder and superconductivity properties in carbohydrate doped MgB_2, increases the J_c density (Kim et al., 2008); the interplay between superconductivity and charge density waves is affected by disorder (Attanasi, 2008); insensitivity of d-wave pairing to disorder in the high temperature cuprate superconductors, increase then decrease from scattering with weak dependence of T_c on n_{imp} impurity in theory as compared to experimental results (Kemper et al., 2009).

In the 2010s, for strongly disordered TiN and NbTiN s-wave superconductors probed by microwave electrodynamics (Driessen, Coumou, Tromp, de Visser, & Klapwijk, 2012), it is mentioned that a decreased T_c with increasing sheet resistance is found for MoGe films by Finkelstein (Finkelstein, 1987); dynamical conductivity across the disorder tuned superconductor–insulator transition, has disorder enhanced absorption in conductivity and expands the quantum critical region (Swanson, Loh, Randeria, & Trivedi, 2014); effects of randomness on T_c in quasi-2D organic superconductors, leads to lowered T_c (Nakhmedov, Alekperov, & Oppermann, 2012).

We note that the interested reader can refer to earlier works involving elementary excitations in solids, quasi-particles, and many-body effects in Pines (1964, 1979), Ginzburg and Kirzhnits (1982), and Kittel (1987) (which includes magnons). Those interested in a more refined electron–phonon interaction model (beyond the two-valued step function model) using the spectral density $\alpha^2(\omega)F(\omega)$, can look at the very thorough review of Corbette (1990), which provides detailed information on the real and imaginary axis analytical continuation relation between, of the solution to the Eliashberg (1960, 1961) equations for the non-weak or strong coupling regimes. It is interesting that with some slight modifications, the strong case for the T_c expression is similar to the weak form, which is treated herein. See Bartolf (2016) for a recent discussion of this and other topics in superconductivity, such as Ginzburg–Landau theory, the two fluid model of Gorter and Casimir, type-II superconductivity with vortices – including work by Abrikosov (2004) and others (Karnaukhov & Shepelev, 2008), and the London theory; Tinkham (1980), Schrieffer (1983) for details on the microscopic approach, as well as Abrikosov, Gorkov, and Dzyaloshinski (1963, 1965) and Abrikosov and Gor'kov (1959a) (electrodynamics using a Matsubara thermodynamic approach for δ, relationship between $\mathbf{j}$ and $\mathbf{A}$ [i.e., Q hitting $\mathbf{A}$], and penetration depth δ), Abrikosov and Gor'kov (1959b) (obtains $T = 0$ electrodynamics equations, with introduction of the $\mathbf{A}$ photonic field Green's function D and the system fermions Green's functions G and F, again finding Q and δ), Abrikosov and Gor'kov (1961) (magnetic type impurities are expected to break the time reversal symmetry of the Cooper pairs, so although this is interesting work, our intent in here is to use materials which do not intentionally have such symmetry breaking). For a review of localized impurity states, see Balatsky, Vekhter, and Zhu (2006).

Here we show that the introduction of disorder through non-alloy impurities could possibly lead to an increase in the superconducting gap $\Delta(T)$. Based upon a second quantized phonon operator, an electron–phonon interaction potential, Matsubara quantum many-body Green's functions employed for phonons and electrons, with determination of the perturbed electron Matsubara quantum many-body Green's from bare values, the perturbed phonon quantum Green's function is employed in an RPA approach, for obtaining permittivity and the superconducting gap $\Delta(T)$.

A microscopic thermodynamic statistical approach, utilizing an energy dispersion relation, averaged particle energy, and entropy, is employed to find critical magnetic field H_c and its slope dH_c/dT, and heat capacity. Finally, the indirect relationship between T_c and gap parameter Δ is shown when obtaining Δ's dependence on the same disorder potential energy quantity as T_c. All of this analysis is done while not necessarily maintaining band structure symmetry at various levels of development.

In what follows, Section 2 reviews the relevant statistical energies from a microscopic thermodynamic statistical perspective, which will be utilized for finding critical magnetic field and its slope, and heat capacity, and other parameters. Section 3 determines the microscopic energy dispersion relation from a Bogoliubov transformation employing a mean field BCS Hamiltonian. Section 4 finds the averaged microscopic energy E formula for calculating the free energy, while Sections 5 and 6 find the microscopic entropy S formula and evaluate it. Section 7 evaluates the microscopic averaged energy E formula. Section 8 determines a microscopic formula for critical magnetic field and evaluates it for a specialized case. Section 9 obtains the microscopic specific heat of the system, whereas Section 10 finds the microscopic slope of the critical magnetic field at T_c. Section 11 relates the disorder potential energy to the gap parameter. Section 12 draws conclusions.

2. RELEVANT STATISTICAL ENERGIES TO USE FOR FINDING CRITICAL MAGNETIC FIELD, HEAT CAPACITY, AND OTHER PARAMETERS

There is a question of what stochastic energies to employ for deriving microscopic parameters of a superconducting system. One might enlist the Helmholtz free energy F because it uses the system energy E, system temperature T, and system entropy S. These are readily associated with statistical and thermodynamic physics quantities which are in the macroscopic sense familiar to us (Reif, 1965),

$$F = E - TS \tag{1}$$

But what about using the Gibbs free energy G, another familiar statistical measure of the system? It is

$$\begin{aligned} G &= E - TS + p\mathcal{V} \\ &= F + p\mathcal{V} \end{aligned} \tag{2}$$

where p is the system pressure and $\mathcal{V}$ its volume. By the reasoning in Kittel (1968), finding an analogy between applied magnetic field H_a and pressure p, and magnetization M and volume $\mathcal{V}$, that is the associations

$$p \leftrightarrow H_a; \qquad \mathcal{V} \leftrightarrow -M \tag{3}$$

we may write G as

$$G = E - TS - H_a M \tag{4}$$

For a non-magnetic system where the magnetizations is zero,

$$M_{non\text{-}mag} = 0 \tag{5}$$

the Gibbs free energy reduces to the free energy,

$$\begin{aligned} G &= \{E - TS - H_a M\}|_{non\text{-}mag} \\ &= E - TS - H_a M_{non\text{-}mag} \\ &= E - TS \\ &= F \end{aligned} \tag{6}$$

Therefore, it is reasonable to focus on obtaining the Helmholtz free energy F, and this will be done in the following sections, requiring us to find E and S.

3. ENERGY DISPERSION RELATION OBTAINED FROM A BOGOLIUBOV TRANSFORMATION ON A MEAN FIELD BCS HAMILTONIAN

Knowing that paired opposite momentum and spin states are associated with a gap parameter $\Delta_{\mathbf{k}}$, and this represents a reduction in energy or energy storage, and $\xi_{\mathbf{k}}$ associated with energy off of the Fermi level, a mean field BCS like Hamiltonian is stated as

$$\begin{aligned}
H_{BCS}^{MF} &= \sum_{\mathbf{k}}\sum_{\sigma}\xi_{\mathbf{k}\sigma}c_{\mathbf{k}\sigma}^{\dagger}c_{\mathbf{k}\sigma} - \sum_{\mathbf{k}}\Delta_{\mathbf{k}}c_{\mathbf{k}\uparrow}^{\dagger}c_{-\mathbf{k}\downarrow}^{\dagger} - \sum_{\mathbf{k}}\Delta_{\mathbf{k}}^{*}c_{-\mathbf{k}\downarrow}c_{\mathbf{k}\uparrow}\\
&= \sum_{\mathbf{k}}\xi_{\mathbf{k}\uparrow}c_{\mathbf{k}\uparrow}^{\dagger}c_{\mathbf{k}\uparrow} + \sum_{\mathbf{k}}\xi_{\mathbf{k}\downarrow}c_{\mathbf{k}\downarrow}^{\dagger}c_{\mathbf{k}\downarrow} - \sum_{\mathbf{k}}\Delta_{\mathbf{k}}c_{\mathbf{k}\uparrow}^{\dagger}c_{-\mathbf{k}\downarrow}^{\dagger} - \sum_{\mathbf{k}}\Delta_{\mathbf{k}}^{*}c_{-\mathbf{k}\downarrow}c_{\mathbf{k}\uparrow}\\
&= \sum_{\mathbf{k}}\xi_{\mathbf{k}\uparrow}c_{\mathbf{k}\uparrow}^{\dagger}c_{\mathbf{k}\uparrow} + \sum_{\mathbf{k}}\xi_{-\mathbf{k}\downarrow}c_{-\mathbf{k}\downarrow}^{\dagger}c_{-\mathbf{k}\downarrow} - \sum_{\mathbf{k}}\Delta_{\mathbf{k}}c_{\mathbf{k}\uparrow}^{\dagger}c_{-\mathbf{k}\downarrow}^{\dagger} - \sum_{\mathbf{k}}\Delta_{\mathbf{k}}^{*}c_{-\mathbf{k}\downarrow}c_{\mathbf{k}\uparrow}\\
&= \sum_{\mathbf{k}}\xi_{\mathbf{k}\uparrow}c_{\mathbf{k}\uparrow}^{\dagger}c_{\mathbf{k}\uparrow} + \sum_{\mathbf{k}}\xi_{-\mathbf{k}\downarrow}\left(1 - c_{-\mathbf{k}\downarrow}c_{-\mathbf{k}\downarrow}^{\dagger}\right) - \sum_{\mathbf{k}}\Delta_{\mathbf{k}}c_{\mathbf{k}\uparrow}^{\dagger}c_{-\mathbf{k}\downarrow}^{\dagger}\\
&\quad - \sum_{\mathbf{k}}\Delta_{\mathbf{k}}^{*}c_{-\mathbf{k}\downarrow}c_{\mathbf{k}\uparrow}
\end{aligned} \tag{7}$$

or in matrix form,

$$\begin{aligned}
H_{BCS}^{MF} &= \sum_{\mathbf{k}}\begin{bmatrix} c_{\mathbf{k}\uparrow}^{\dagger} & 0 \end{bmatrix}\begin{bmatrix} \xi_{\mathbf{k}\uparrow} & 0 \\ 0 & 0 \end{bmatrix}\begin{bmatrix} c_{\mathbf{k}\uparrow} \\ 0 \end{bmatrix}\\
&\quad + \sum_{\mathbf{k}}\begin{bmatrix} 0 & c_{-\mathbf{k}\downarrow} \end{bmatrix}\begin{bmatrix} 0 & 0 \\ 0 & -\xi_{-\mathbf{k}\downarrow} \end{bmatrix}\begin{bmatrix} 0 \\ c_{-\mathbf{k}\downarrow}^{\dagger} \end{bmatrix} + \sum_{\mathbf{k}}\xi_{-\mathbf{k}\downarrow}\\
&\quad + \sum_{\mathbf{k}}\begin{bmatrix} c_{\mathbf{k}\uparrow}^{\dagger} & 0 \end{bmatrix}\begin{bmatrix} 0 & -\Delta_{\mathbf{k}} \\ 0 & 0 \end{bmatrix}\begin{bmatrix} 0 \\ c_{-\mathbf{k}\downarrow}^{\dagger} \end{bmatrix}\\
&\quad - \sum_{\mathbf{k}}\left\{\begin{bmatrix} c_{\mathbf{k}\uparrow}^{\dagger} & 0 \end{bmatrix}\begin{bmatrix} 0 & -\Delta_{\mathbf{k}} \\ 0 & 0 \end{bmatrix}\begin{bmatrix} 0 \\ c_{-\mathbf{k}\downarrow}^{\dagger} \end{bmatrix}\right\}^{\dagger}\\
&= \sum_{\mathbf{k}}\begin{bmatrix} c_{\mathbf{k}\uparrow}^{\dagger} & c_{-\mathbf{k}\downarrow} \end{bmatrix}\begin{bmatrix} \xi_{\mathbf{k}\uparrow} & 0 \\ 0 & 0 \end{bmatrix}\begin{bmatrix} c_{\mathbf{k}\uparrow} \\ c_{-\mathbf{k}\downarrow}^{\dagger} \end{bmatrix}\\
&\quad + \sum_{\mathbf{k}}\begin{bmatrix} c_{\mathbf{k}\uparrow}^{\dagger} & c_{-\mathbf{k}\downarrow} \end{bmatrix}\begin{bmatrix} 0 & 0 \\ 0 & -\xi_{-\mathbf{k}\downarrow} \end{bmatrix}\begin{bmatrix} c_{\mathbf{k}\uparrow} \\ c_{-\mathbf{k}\downarrow}^{\dagger} \end{bmatrix} + \sum_{\mathbf{k}}\xi_{-\mathbf{k}\downarrow}\\
&\quad + \sum_{\mathbf{k}}\begin{bmatrix} c_{\mathbf{k}\uparrow}^{\dagger} & c_{-\mathbf{k}\downarrow} \end{bmatrix}\begin{bmatrix} 0 & -\Delta_{\mathbf{k}} \\ 0 & 0 \end{bmatrix}\begin{bmatrix} c_{\mathbf{k}\uparrow} \\ c_{-\mathbf{k}\downarrow}^{\dagger} \end{bmatrix}
\end{aligned}$$

$$
\begin{aligned}
&-\sum_{\mathbf{k}}\left\{\begin{bmatrix} c_{\mathbf{k}\uparrow}^{\dagger} & c_{-\mathbf{k}\downarrow} \end{bmatrix}\begin{bmatrix} 0 & -\Delta_{\mathbf{k}} \\ 0 & 0 \end{bmatrix}\begin{bmatrix} c_{\mathbf{k}\uparrow} \\ c_{-\mathbf{k}\downarrow}^{\dagger} \end{bmatrix}\right\}^{\dagger} \\
&=\sum_{\mathbf{k}}\begin{bmatrix} c_{\mathbf{k}\uparrow}^{\dagger} & c_{-\mathbf{k}\downarrow} \end{bmatrix}\begin{bmatrix} \xi_{\mathbf{k}\uparrow} & 0 \\ 0 & -\xi_{-\mathbf{k}\downarrow} \end{bmatrix}\begin{bmatrix} c_{\mathbf{k}\uparrow} \\ c_{-\mathbf{k}\downarrow}^{\dagger} \end{bmatrix}+\sum_{\mathbf{k}}\xi_{-\mathbf{k}\downarrow} \\
&\quad+\sum_{\mathbf{k}}\begin{bmatrix} c_{\mathbf{k}\uparrow}^{\dagger} & c_{-\mathbf{k}\downarrow} \end{bmatrix}\begin{bmatrix} 0 & -\Delta_{\mathbf{k}} \\ 0 & 0 \end{bmatrix}\begin{bmatrix} c_{\mathbf{k}\uparrow} \\ c_{-\mathbf{k}\downarrow}^{\dagger} \end{bmatrix} \\
&\quad-\sum_{\mathbf{k}}\begin{bmatrix} c_{\mathbf{k}\uparrow}^{\dagger} & c_{-\mathbf{k}\downarrow} \end{bmatrix}\begin{bmatrix} 0 & 0 \\ -\Delta_{\mathbf{k}}^{*} & 0 \end{bmatrix}\begin{bmatrix} c_{\mathbf{k}\uparrow} \\ c_{-\mathbf{k}\downarrow}^{\dagger} \end{bmatrix} \\
&=\sum_{\mathbf{k}}\begin{bmatrix} c_{\mathbf{k}\uparrow}^{\dagger} & c_{-\mathbf{k}\downarrow} \end{bmatrix}\begin{bmatrix} \xi_{\mathbf{k}\uparrow} & 0 \\ 0 & -\xi_{-\mathbf{k}\downarrow} \end{bmatrix}\begin{bmatrix} c_{\mathbf{k}\uparrow} \\ c_{-\mathbf{k}\downarrow}^{\dagger} \end{bmatrix}+\sum_{\mathbf{k}}\xi_{-\mathbf{k}\downarrow} \\
&\quad+\sum_{\mathbf{k}}\begin{bmatrix} c_{\mathbf{k}\uparrow}^{\dagger} & c_{-\mathbf{k}\downarrow} \end{bmatrix}\begin{bmatrix} 0 & -\Delta_{\mathbf{k}} \\ -\Delta_{\mathbf{k}}^{*} & 0 \end{bmatrix}\begin{bmatrix} c_{\mathbf{k}\uparrow} \\ c_{-\mathbf{k}\downarrow}^{\dagger} \end{bmatrix} \\
&=\sum_{\mathbf{k}}\begin{bmatrix} c_{\mathbf{k}\uparrow}^{\dagger} & c_{-\mathbf{k}\downarrow} \end{bmatrix}\begin{bmatrix} \xi_{\mathbf{k}\uparrow} & -\Delta_{\mathbf{k}} \\ -\Delta_{\mathbf{k}}^{*} & -\xi_{-\mathbf{k}\downarrow} \end{bmatrix}\begin{bmatrix} c_{\mathbf{k}\uparrow} \\ c_{-\mathbf{k}\downarrow}^{\dagger} \end{bmatrix}+\sum_{\mathbf{k}}\xi_{-\mathbf{k}\downarrow} \\
&=\sum_{\mathbf{k}}A_{\mathbf{k}\uparrow\downarrow}^{\dagger}\begin{bmatrix} \xi_{\mathbf{k}\uparrow} & -\Delta_{\mathbf{k}} \\ -\Delta_{\mathbf{k}}^{*} & -\xi_{-\mathbf{k}\downarrow} \end{bmatrix}A_{\mathbf{k}\uparrow\downarrow}+\sum_{\mathbf{k}}\xi_{-\mathbf{k}\downarrow}
\end{aligned} \tag{8}
$$

where column c-matrix $A_{\mathbf{k}\uparrow\downarrow}$ is given by

$$
A_{\mathbf{k}\uparrow\downarrow}=\begin{bmatrix} c_{\mathbf{k}\uparrow} \\ c_{-\mathbf{k}\downarrow}^{\dagger} \end{bmatrix} \tag{9}
$$

and the second sum in (8) is a constant determined as

$$
\begin{aligned}
\sum_{\mathbf{k}}\xi_{-\mathbf{k}\downarrow}&=\sum_{\mathbf{k}}\xi_{\mathbf{k}\downarrow}=\sum_{\mathbf{k}}(\varepsilon_{\mathbf{k}\downarrow}-\varepsilon_{F\mathbf{k}})=\frac{1}{2}E^{(0)}-\varepsilon_F\sum_{\mathbf{k}}1 \\
&=\frac{1}{2}E^{(0)}-\varepsilon_F\frac{N}{2}=-\frac{1}{3}E^{(0)}
\end{aligned} \tag{10}
$$

Here as one progresses to an actual value, spherical energy (in 3D) or circular energy (in 2D), and Fermi surface symmetry are assumed.

There is a way using a 2×2 matrix transformation, because of the final form of (8), to diagonalize it, by using a Bogoliubov procedure mapping the c 2nd quantized operators into γ 2nd quantized operators,

$$B_{\mathbf{k}\uparrow\downarrow} = \begin{bmatrix} \gamma_{\mathbf{k}\uparrow} \\ \gamma^{\dagger}_{\mathbf{k}\downarrow} \end{bmatrix} = \begin{bmatrix} u^{*}_{\mathbf{k}} & v_{\mathbf{k}} \\ -v^{*}_{\mathbf{k}} & u_{\mathbf{k}} \end{bmatrix} \begin{bmatrix} c_{\mathbf{k}\uparrow} \\ c^{\dagger}_{-\mathbf{k}\downarrow} \end{bmatrix} = M_{BA} A_{\mathbf{k}\uparrow\downarrow};$$
$$M_{BA} = \begin{bmatrix} u^{*}_{\mathbf{k}} & v_{\mathbf{k}} \\ -v^{*}_{\mathbf{k}} & u_{\mathbf{k}} \end{bmatrix} \tag{11}$$

The result is

$$H^{MF}_{BCS} = \sum_{\mathbf{k}} B^{\dagger}_{\mathbf{k}\uparrow\downarrow} H^{Bogol}_{\mathbf{k}} B_{\mathbf{k}\uparrow\downarrow} + \sum_{\mathbf{k}} \xi_{-\mathbf{k}\downarrow} \tag{12}$$

where

$$H^{Bogol}_{\mathbf{k}} = M_{BA} H_{\mathbf{k}} M^{\dagger}_{BA} = \begin{bmatrix} E_{\mathbf{k}} & 0 \\ 0 & \tilde{E}_{\mathbf{k}} \end{bmatrix}; \qquad H_{\mathbf{k}} = \begin{bmatrix} \xi_{\mathbf{k}\uparrow} & -\Delta_{\mathbf{k}} \\ -\Delta^{*}_{\mathbf{k}} & -\xi_{-\mathbf{k}\downarrow} \end{bmatrix};$$
$$M^{-1}_{BA} = [M_{BA}]^{-1} = M^{\dagger}_{BA} \tag{13}$$

with

$$E_{\mathbf{k}} = \xi_{\mathbf{k}\uparrow} |u_{\mathbf{k}}|^2 - \xi_{-\mathbf{k}\downarrow} |v_{\mathbf{k}}|^2 - \Delta_{\mathbf{k}} u^{*}_{\mathbf{k}} v^{*}_{\mathbf{k}} - \Delta^{*}_{\mathbf{k}} u_{\mathbf{k}} v_{\mathbf{k}} \tag{14}$$
$$\tilde{E}_{\mathbf{k}} = \xi_{\mathbf{k}\uparrow} |v_{\mathbf{k}}|^2 - \xi_{-\mathbf{k}\downarrow} |u_{\mathbf{k}}|^2 + \Delta_{\mathbf{k}} u^{*}_{\mathbf{k}} v^{*}_{\mathbf{k}} + \Delta^{*}_{\mathbf{k}} u_{\mathbf{k}} v_{\mathbf{k}} \tag{15}$$

so that the diagonal element sum is

$$E_{\mathbf{k}} + \tilde{E}_{\mathbf{k}} = \xi_{\mathbf{k}\uparrow} - \xi_{-\mathbf{k}\downarrow} \tag{16}$$

using the $u_{\mathbf{k}}$ and $v_{\mathbf{k}}$ normalization

$$|u_{\mathbf{k}}|^2 + |v_{\mathbf{k}}|^2 = 1 \tag{17}$$

Generally, inspecting (14) and (15), we see that $\tilde{E}_{\mathbf{k}}$ is not equal to $-E_{\mathbf{k}}$, because electron energy may be both spin and momentum surface dependent. Therefore, the Bogoliubov Hamiltonian is written as

$$H^{Bogol}_{\mathbf{k}} = \begin{bmatrix} E_{\mathbf{k}} & 0 \\ 0 & -E_{\mathbf{k}} + (\xi_{\mathbf{k}\uparrow} - \xi_{-\mathbf{k}\downarrow}) \end{bmatrix} \tag{18}$$

Only when the energy loses spin dependence, and the Fermi surface has enough symmetry that its value looks the same at **k** and −**k**, can we write

$$\xi_{\mathbf{k}\uparrow} = \xi_{-\mathbf{k}\downarrow} = \xi_{\mathbf{k}} \tag{19}$$

Spherical symmetry in 3D, circular symmetry in 2D, further reduces the energy to be only momentum magnitude dependent, $\xi_{\mathbf{k}} = \xi_k$.

Placing (18) and the Bogoliubons into (12), enlisting the diagonalized form in (13), mean field BCS-like Hamiltonian becomes

$$\begin{aligned} H_{BCS}^{MF} &= \sum_{\mathbf{k}} \begin{bmatrix} \gamma_{\mathbf{k}\uparrow}^{\dagger} & \gamma_{-\mathbf{k}\downarrow} \end{bmatrix} \begin{bmatrix} E_{\mathbf{k}} & 0 \\ 0 & \tilde{E}_{\mathbf{k}} \end{bmatrix} \begin{bmatrix} \gamma_{\mathbf{k}\uparrow} \\ \gamma_{-\mathbf{k}\downarrow}^{\dagger} \end{bmatrix} + \sum_{\mathbf{k}} \xi_{-\mathbf{k}\downarrow} \\ &= \sum_{\mathbf{k}} (E_{\mathbf{k}} \gamma_{\mathbf{k}\uparrow}^{\dagger} \gamma_{\mathbf{k}\uparrow} + \tilde{E}_{\mathbf{k}} \gamma_{-\mathbf{k}\downarrow} \gamma_{-\mathbf{k}\downarrow}^{\dagger}) + \sum_{\mathbf{k}} \xi_{-\mathbf{k}\downarrow} \\ &= \sum_{\mathbf{k}} E_{\mathbf{k}} \gamma_{\mathbf{k}\uparrow}^{\dagger} \gamma_{\mathbf{k}\uparrow} + \sum_{-\mathbf{k}} \tilde{E}_{-\mathbf{k}} \gamma_{\mathbf{k}\downarrow} \gamma_{\mathbf{k}\downarrow}^{\dagger} + \sum_{\mathbf{k}} \xi_{-\mathbf{k}\downarrow} \\ &= \sum_{\mathbf{k}} (E_{\mathbf{k}} \gamma_{\mathbf{k}\uparrow}^{\dagger} \gamma_{\mathbf{k}\uparrow} + \tilde{E}_{-\mathbf{k}} \gamma_{\mathbf{k}\downarrow} \gamma_{\mathbf{k}\downarrow}^{\dagger}) + \sum_{\mathbf{k}} \xi_{-\mathbf{k}\downarrow} \\ &= \sum_{\mathbf{k}} (E_{\mathbf{k}} \gamma_{\mathbf{k}\uparrow}^{\dagger} \gamma_{\mathbf{k}\uparrow} + [E_{-\mathbf{k}} + (\xi_{-\mathbf{k}\uparrow} - \xi_{\mathbf{k}\downarrow})] \gamma_{\mathbf{k}\downarrow}^{\dagger} \gamma_{\mathbf{k}\downarrow}) \\ &\quad + \sum_{\mathbf{k}} ([\xi_{-\mathbf{k}\uparrow} - \xi_{\mathbf{k}\downarrow}] - E_{-\mathbf{k}}) + \sum_{\mathbf{k}} \xi_{-\mathbf{k}\downarrow} \end{aligned} \tag{20}$$

Second and third summation terms will be some constant non-2nd quantized operator values.

Properties of $u_{\mathbf{k}}$, $v_{\mathbf{k}}$, and $E_{\mathbf{k}}$ can be pinned down by solving the transformation in (13) using unitary matrix M_{BA}. Equating $H_{\mathbf{k}}^{Bogol}$ to the transformed $H_{\mathbf{k}}$,

$$\begin{bmatrix} E_{\mathbf{k}} & 0 \\ 0 & \tilde{E}_{\mathbf{k}} \end{bmatrix} = \begin{bmatrix} u_{\mathbf{k}}^* & v_{\mathbf{k}} \\ -v_{\mathbf{k}}^* & u_{\mathbf{k}} \end{bmatrix} \begin{bmatrix} \xi_{\mathbf{k}\uparrow} & -\Delta_{\mathbf{k}} \\ -\Delta_{\mathbf{k}}^* & -\xi_{-\mathbf{k}\downarrow} \end{bmatrix} \begin{bmatrix} u_{\mathbf{k}} & -v_{\mathbf{k}} \\ v_{\mathbf{k}}^* & u_{\mathbf{k}}^* \end{bmatrix} \tag{21}$$

For the off-diagonal elements of $H_{\mathbf{k}}^{Bogol}$,

$$-(H_{\mathbf{k}}^{Bogol})_{21} = -(H_{\mathbf{k}}^{Bogol})_{12}^{\bullet} = (\xi_{\mathbf{k}\uparrow} + \xi_{-\mathbf{k}\downarrow}) v_{\mathbf{k}}^* u_{\mathbf{k}} + \Delta_{\mathbf{k}}^* (u_{\mathbf{k}})^2 - \Delta_{\mathbf{k}} (v_{\mathbf{k}}^*)^2 = 0 \tag{22}$$

Combined with (14)–(16), (22) yields the solution for the parameters. The last result in (20) for the mean field BCS-like Hamiltonian may be recast in a slightly different form as

$$H_{BCS}^{MF} = \sum_{\mathbf{k}} (E_{\mathbf{k}} \gamma_{\mathbf{k}\uparrow}^{\dagger} \gamma_{\mathbf{k}\uparrow} + E'_{-\mathbf{k}} \gamma_{\mathbf{k}\downarrow}^{\dagger} \gamma_{\mathbf{k}\downarrow}) + \sum_{\mathbf{k}} E_{c\mathbf{k}} \tag{23}$$

where $E_{\mathbf{k}}$ is still given by (14), but $E'_{\mathbf{k}}$ by

$$E'_{\mathbf{k}} = \xi_{-\mathbf{k}\downarrow} |u_{\mathbf{k}}|^2 - \xi_{\mathbf{k}\uparrow} |v_{\mathbf{k}}|^2 - \Delta_{\mathbf{k}} u_{\mathbf{k}}^* v_{\mathbf{k}}^* - \Delta_{\mathbf{k}}^* u_{\mathbf{k}} v_{\mathbf{k}} \tag{24}$$

$$E_{c\mathbf{k}} = (\xi_{-\mathbf{k}\downarrow} + \xi_{\mathbf{k}\uparrow})|v_{\mathbf{k}}|^2 + \Delta_{\mathbf{k}} u_{\mathbf{k}}^* v_{\mathbf{k}}^* + \Delta_{\mathbf{k}}^* u_{\mathbf{k}} v_{\mathbf{k}} \tag{25}$$

Magnitudes of the $u_{\mathbf{k}}$, $v_{\mathbf{k}}$ are given by

$$|u_{\mathbf{k}}|^2 = \frac{1}{2}\left(1 + \frac{\xi_{\mathbf{k}a}}{E_{\mathbf{k}a}}\right); \qquad |v_{\mathbf{k}}|^2 = \frac{1}{2}\left(1 - \frac{\xi_{\mathbf{k}a}}{E_{\mathbf{k}a}}\right) \tag{26}$$

where the averaged electron Fermion and quasi-particle energy are

$$\xi_{\mathbf{k}a} = \frac{\xi_{\mathbf{k}\uparrow} + \xi_{-\mathbf{k}\downarrow}}{2}; \qquad E_{\mathbf{k}a} = \frac{E_{\mathbf{k}} + E'_{\mathbf{k}}}{2} \tag{27}$$

$E_{\mathbf{k}a}$ is by (27),

$$E_{\mathbf{k}a} = \xi_{\mathbf{k}a}\left(|u_{\mathbf{k}}|^2 - |v_{\mathbf{k}}|^2\right) - \Delta_{\mathbf{k}} u_{\mathbf{k}}^* v_{\mathbf{k}}^* - \Delta_{\mathbf{k}}^* u_{\mathbf{k}} v_{\mathbf{k}} \tag{28}$$

which has two branches,

$$E_{\mathbf{k}a} = \pm\sqrt{\xi_{\mathbf{k}a}^2 + |\Delta_{\mathbf{k}}|^2} \tag{29}$$

The dispersion equation (29) arose from addressing the complex nature of the parameters, and also leads to the diagonal Bogoliubons in the Hamiltonian of (23). Lets see how, writing

$$u_{\mathbf{k}} = |u_{\mathbf{k}}|e^{i\vartheta_u}; \qquad v_{\mathbf{k}} = |v_{\mathbf{k}}|e^{i\vartheta_v}; \qquad \Delta_{\mathbf{k}} = |\Delta_{\mathbf{k}}|e^{i\vartheta_\Delta} \tag{30}$$

Inserting (23) into the off-diagonal constraint equation (22), we arrive at two conditions

$$2\xi_{\mathbf{k}a}|u_{\mathbf{k}}||v_{\mathbf{k}}| + |\Delta_{\mathbf{k}}|\left(2|u_{\mathbf{k}}|^2 - 1\right)\cos(\vartheta_D) = 0 \tag{31a}$$

$$|\Delta_{\mathbf{k}}|\sin(\vartheta_D) = 0 \tag{31b}$$

where the difference angle is the difference between the gap angle and the sum of the other two parameter angles:

$$\vartheta_D = \vartheta_\Delta - (\vartheta_u + \vartheta_v) \tag{32}$$

The second (31) equation forces ϑ_D to be held to

$$\vartheta_D = n\pi; \quad n = 0, \pm 1, \pm 2, \cdots \tag{33}$$

leading to the diagonal energy values of $E_{\mathbf{k}}$ and $E'_{\mathbf{k}}$,

$$E_{\mathbf{k}} = \frac{\xi_{\mathbf{k}\uparrow} - \xi_{-\mathbf{k}\downarrow}}{2} + E_{\mathbf{k}a}; \qquad E'_{\mathbf{k}} = E_{\mathbf{k}a} - \frac{\xi_{\mathbf{k}\uparrow} - \xi_{-\mathbf{k}\downarrow}}{2} \tag{34}$$

When the system develops enough symmetry that (19) is true, (27) and (29) reduce to

$$\xi_{\mathbf{k}a} \rightarrow \xi_{\mathbf{k}}; \qquad E_{\mathbf{k}a} \rightarrow E_{\mathbf{k}} = \pm\sqrt{\xi_{\mathbf{k}}^2 + |\Delta_{\mathbf{k}}|^2} \tag{35}$$

the familiar forms.

4. AVERAGED ENERGY E TO BE USED IN THE FREE ENERGY CONSTRUCTION

The free energy F which relates to macroscopic statistical thermodynamics, as utilized in (1) and (2), requires obtaining the energy E which must be a statistical averaged value found from the Hamiltonian in the first line of (8), where the gap parameter is

$$\Delta_{\mathbf{k}} = \sum_{\mathbf{k}'} V'_{\mathbf{k}\mathbf{k}'} \langle c_{-\mathbf{k}'\downarrow} c_{\mathbf{k}\uparrow} \rangle \tag{36}$$

with $V'_{\mathbf{k}\mathbf{k}'}$ found by using,

$$\begin{aligned} V'^{RPA,ren}_{eff}(\mathbf{k}, \mathbf{k}'; ik_n, ik'_n) = {} & V'^{RPA,CP}_{eff}(\mathbf{k}, \mathbf{k}'; ik_n, ik'_n) \\ & + V'^{RPA,imp}_{eff}(\mathbf{k}, \mathbf{k}'; ik_n, ik'_n) \end{aligned} \tag{37}$$

by extension off of the diagonal form, using Coulomb-phonon effective potential and RPA impurity potential energies. Dropping the 4th momentum space variables in each potential energy, using the reasoning that we are just looking at small energies, leads us to the formula

$$V'_{\mathbf{k}\mathbf{k}'} = V'^{RPA,ren}_{eff}(\mathbf{k}, \mathbf{k}') = V'^{RPA,CP}_{eff}(\mathbf{k}, \mathbf{k}') + V'^{RPA,imp}_{eff}(\mathbf{k}, \mathbf{k}') \tag{38}$$

The BCS wavefunction $|\psi_{BCS}\rangle$ is expressible as the product of the $F_{\mathbf{k}}$ operator and the vacuum state

$$|\psi_{BCS}\rangle = \left(\prod_{\mathbf{k}} F_{\mathbf{k}}\right)|0\rangle = \prod_{\mathbf{k}} (F_{\mathbf{k}}|0_{\mathbf{k}}\rangle); \qquad F_{\mathbf{k}} = u_{\mathbf{k}} + v_{\mathbf{k}} c^{\dagger}_{\mathbf{k}\uparrow} c^{\dagger}_{-\mathbf{k}\downarrow} \tag{39}$$

Using $|\psi_{BCS}\rangle$ from (39), an energy quantity which accounts for all of the momentum states in the superconductor system, creates a statistical based number, by hitting H^{MF}_{BCS} with the bra $\langle\psi_{BCS}|$ and ket $|\psi_{BCS}\rangle$ system states

$$E = \langle\psi_{BCS}|H^{MF}_{BCS}|\psi_{BCS}\rangle$$

$$
\begin{aligned}
&= \left[\prod_{\mathbf{k}'} (\langle 0_{\mathbf{k}'}| F^\dagger_{\mathbf{k}'})\right] \left\{ \sum_{\bar{\mathbf{k}}} (\xi_{\bar{\mathbf{k}}\uparrow} n_{\bar{\mathbf{k}}\uparrow} + \xi_{-\bar{\mathbf{k}}\downarrow} n_{\bar{\mathbf{k}}\downarrow}) \right\} \left[\prod_{\mathbf{k}} (F_{\mathbf{k}} |0_{\mathbf{k}}\rangle)\right] \\
&+ \left[\prod_{\mathbf{k}'} (\langle 0_{\mathbf{k}'}| F^\dagger_{\mathbf{k}'})\right] \left\{ -\sum_{\bar{\mathbf{k}}} \Delta_{\bar{\mathbf{k}}} c^\dagger_{\bar{\mathbf{k}}\uparrow} c^\dagger_{-\bar{\mathbf{k}}\downarrow} \right\} \left[\prod_{\mathbf{k}} (F_{\mathbf{k}} |0_{\mathbf{k}}\rangle)\right] \\
&+ \left\{ \left[\prod_{\mathbf{k}'} (\langle 0_{\mathbf{k}'}| F^\dagger_{\mathbf{k}'})\right] \left\{ -\sum_{\bar{\mathbf{k}}} \Delta_{\bar{\mathbf{k}}} c^\dagger_{\bar{\mathbf{k}}\uparrow} c^\dagger_{-\bar{\mathbf{k}}\downarrow} \right\} \left[\prod_{\mathbf{k}} (F_{\mathbf{k}} |0_{\mathbf{k}}\rangle)\right] \right\}^\dagger \\
&= \sum_{\bar{\mathbf{k}}} \left[\prod_{\mathbf{k}' \neq \bar{\mathbf{k}}} (\langle 0_{\mathbf{k}'}| F^\dagger_{\mathbf{k}'})\right] \left[\prod_{\mathbf{k} \neq \bar{\mathbf{k}}} (F_{\mathbf{k}} |0_{\mathbf{k}}\rangle)\right] \langle 0_{\bar{\mathbf{k}}}| F^\dagger_{\bar{\mathbf{k}}} (\xi_{\bar{\mathbf{k}}\uparrow} n_{\mathbf{k}\uparrow} + \xi_{-\bar{\mathbf{k}}\downarrow} n_{\bar{\mathbf{k}}\downarrow}) F_{\bar{\mathbf{k}}} |0_{\bar{\mathbf{k}}}\rangle \\
&- \sum_{\bar{\mathbf{k}}} \Delta_{\bar{\mathbf{k}}} \left[\prod_{\mathbf{k}' \neq \bar{\mathbf{k}}} (\langle 0_{\mathbf{k}'}| F^\dagger_{\mathbf{k}'})\right] \left[\prod_{\mathbf{k} \neq \bar{\mathbf{k}}} (F_{\mathbf{k}} |0_{\mathbf{k}}\rangle)\right] \langle 0_{\bar{\mathbf{k}}}| F^\dagger c^\dagger_{\bar{\mathbf{k}}\uparrow} c^\dagger_{-\bar{\mathbf{k}}\downarrow} F_{\bar{\mathbf{k}}} |0_{\bar{\mathbf{k}}}\rangle \\
&+ \{2nd\ term\}^\dagger
\end{aligned}
\tag{40}
$$

Eq. (40) further reduces,

$$
\begin{aligned}
E &= \sum_{\bar{\mathbf{k}}} \left[\prod_{\mathbf{k} \neq \bar{\mathbf{k}}} (\langle 0_{\mathbf{k}}| F^\dagger_{\mathbf{k}} F_{\mathbf{k}} |0_{\mathbf{k}}\rangle)\right] (\xi_{\bar{\mathbf{k}}\uparrow} + \xi_{-\bar{\mathbf{k}}\downarrow}) \langle 0_{\bar{\mathbf{k}}}| F^\dagger_{\bar{\mathbf{k}}} n_{\bar{\mathbf{k}}\uparrow} F_{\bar{\mathbf{k}}} |0_{\bar{\mathbf{k}}}\rangle \\
&- \sum_{\bar{\mathbf{k}}} \Delta_{\bar{\mathbf{k}}} \left[\prod_{\mathbf{k} \neq \bar{\mathbf{k}}} (\langle 0_{\mathbf{k}}| F^\dagger_{\mathbf{k}} F_{\mathbf{k}} |0_{\mathbf{k}}\rangle)\right] \langle 0_{\bar{\mathbf{k}}}| F^\dagger_{\bar{\mathbf{k}}} c^\dagger_{\bar{\mathbf{k}}\uparrow} c^\dagger_{-\bar{\mathbf{k}}\downarrow} (u_{\bar{\mathbf{k}}} + v_{\bar{\mathbf{k}}} c^\dagger_{\bar{\mathbf{k}}\uparrow} c^\dagger_{-\bar{\mathbf{k}}\downarrow}) |0_{\bar{\mathbf{k}}}\rangle \\
&+ \{2nd\ term\}^\dagger \\
&= \sum_{\bar{\mathbf{k}}} 1 \cdot (\xi_{\bar{\mathbf{k}}\uparrow} + \xi_{-\bar{\mathbf{k}}\downarrow}) \langle 0_{\bar{\mathbf{k}}}| F^\dagger_{\bar{\mathbf{k}}} c^\dagger_{\bar{\mathbf{k}}\uparrow} c_{\bar{\mathbf{k}}\uparrow} (u_{\bar{\mathbf{k}}} + v_{\bar{\mathbf{k}}} c^\dagger_{\bar{\mathbf{k}}\uparrow} c^\dagger_{-\bar{\mathbf{k}}\downarrow}) |0_{\bar{\mathbf{k}}}\rangle \\
&- \sum_{\bar{\mathbf{k}}} \Delta_{\bar{\mathbf{k}}} \cdot 1 \cdot \langle 0_{\bar{\mathbf{k}}}| F^\dagger_{\bar{\mathbf{k}}} c^\dagger_{\bar{\mathbf{k}}\uparrow} (c^\dagger_{-\bar{\mathbf{k}}\downarrow} u_{\bar{\mathbf{k}}} + v_{\bar{\mathbf{k}}} c^\dagger_{-\bar{\mathbf{k}}\downarrow} c^\dagger_{\bar{\mathbf{k}}\uparrow} c^\dagger_{-\bar{\mathbf{k}}\downarrow}) |0_{\bar{\mathbf{k}}}\rangle + \{2nd\ term\}^\dagger \\
&= \sum_{\bar{\mathbf{k}}} (\xi_{\bar{\mathbf{k}}\uparrow} + \xi_{-\bar{\mathbf{k}}\downarrow}) \langle 0_{\bar{\mathbf{k}}}| F^\dagger_{\bar{\mathbf{k}}} (u_{\bar{\mathbf{k}}} c^\dagger_{\bar{\mathbf{k}}\uparrow} c_{\bar{\mathbf{k}}\uparrow} |0_{\bar{\mathbf{k}}}\rangle + v_{\bar{\mathbf{k}}} c^\dagger_{\bar{\mathbf{k}}\uparrow} c_{\bar{\mathbf{k}}\uparrow} c^\dagger_{\bar{\mathbf{k}}\uparrow} c^\dagger_{-\bar{\mathbf{k}}\downarrow} |0_{\bar{\mathbf{k}}}\rangle) \\
&- \sum_{\bar{\mathbf{k}}} \Delta_{\bar{\mathbf{k}}} \langle 0_{\bar{\mathbf{k}}}| F^\dagger_{\bar{\mathbf{k}}} c^\dagger_{\bar{\mathbf{k}}\uparrow} (c^\dagger_{-\bar{\mathbf{k}}\downarrow} u_{\bar{\mathbf{k}}} |0_{\bar{\mathbf{k}}}\rangle - v_{\bar{\mathbf{k}}} c^\dagger_{\bar{\mathbf{k}}\uparrow} c^\dagger_{-\bar{\mathbf{k}}\downarrow} c^\dagger_{-\bar{\mathbf{k}}\downarrow} |0_{\bar{\mathbf{k}}}\rangle) + \{2nd\ term\}^\dagger \\
&= \sum_{\bar{\mathbf{k}}} (\xi_{\bar{\mathbf{k}}\uparrow} + \xi_{-\bar{\mathbf{k}}\downarrow}) \langle 0_{\bar{\mathbf{k}}}| F^\dagger_{\bar{\mathbf{k}}} (u_{\bar{\mathbf{k}}} c^\dagger_{\bar{\mathbf{k}}\uparrow} \cdot 0 + v_{\bar{\mathbf{k}}} |1_{\bar{\mathbf{k}}\uparrow} 1_{-\bar{\mathbf{k}}\downarrow}\rangle) \\
&- \sum_{\bar{\mathbf{k}}} \Delta_{\bar{\mathbf{k}}} \langle 0_{\bar{\mathbf{k}}}| F^\dagger_{\bar{\mathbf{k}}} c^\dagger_{\bar{\mathbf{k}}\uparrow} (c^\dagger_{-\bar{\mathbf{k}}\downarrow} u_{\bar{\mathbf{k}}} |0_{\bar{\mathbf{k}}}\rangle - v_{\bar{\mathbf{k}}} c^\dagger_{\bar{\mathbf{k}}\uparrow} \cdot 0) + \{2nd\ term\}^\dagger
\end{aligned}
\tag{41a}
$$

$$
E = \sum_{\bar{\mathbf{k}}} (\xi_{\bar{\mathbf{k}}\uparrow} + \xi_{-\bar{\mathbf{k}}\downarrow}) \langle 0_{\bar{\mathbf{k}}}| (u^*_{\bar{\mathbf{k}}} + v^*_{\bar{\mathbf{k}}} c_{-\bar{\mathbf{k}}\downarrow} c_{\bar{\mathbf{k}}\uparrow}) |1_{\bar{\mathbf{k}}\uparrow} 1_{-\bar{\mathbf{k}}\downarrow}\rangle v_{\bar{\mathbf{k}}}
$$

$$-\sum_{\bar{\mathbf{k}}} \Delta_{\bar{\mathbf{k}}} u_{\bar{\mathbf{k}}} \langle 0_{\bar{\mathbf{k}}} | \left(u_{\bar{\mathbf{k}}}^* + v_{\bar{\mathbf{k}}}^* c_{-\bar{\mathbf{k}}\downarrow} c_{\bar{\mathbf{k}}\uparrow} \right) c_{\bar{\mathbf{k}}\uparrow}^\dagger c_{-\bar{\mathbf{k}}\downarrow}^\dagger | 0_{\bar{\mathbf{k}}} \rangle + \{2nd\ term\}^\dagger$$

$$= \sum_{\bar{\mathbf{k}}} (\xi_{\bar{\mathbf{k}}\uparrow} + \xi_{-\bar{\mathbf{k}}\downarrow}) \left(u_{\bar{\mathbf{k}}}^* v_{\bar{\mathbf{k}}} \langle 0_{\bar{\mathbf{k}}} | 1_{\bar{\mathbf{k}}\uparrow} 1_{-\bar{\mathbf{k}}\downarrow} \rangle + v_{\bar{\mathbf{k}}}^* v_{\bar{\mathbf{k}}} \langle 0_{\bar{\mathbf{k}}} | c_{-\bar{\mathbf{k}}\downarrow} c_{\bar{\mathbf{k}}\uparrow} | 1_{\bar{\mathbf{k}}\uparrow} 1_{-\bar{\mathbf{k}}\downarrow} \rangle \right)$$

$$-\sum_{\bar{\mathbf{k}}} \Delta_{\bar{\mathbf{k}}} u_{\bar{\mathbf{k}}} \langle 0_{\bar{\mathbf{k}}} | \left(u_{\bar{\mathbf{k}}}^* c_{\bar{\mathbf{k}}\uparrow}^\dagger c_{-\bar{\mathbf{k}}\downarrow}^\dagger | 0_{\bar{\mathbf{k}}} \rangle + v_{\bar{\mathbf{k}}}^* c_{-\bar{\mathbf{k}}\downarrow} c_{\bar{\mathbf{k}}\uparrow} c_{\bar{\mathbf{k}}\uparrow}^\dagger c_{-\bar{\mathbf{k}}\downarrow}^\dagger | 0_{\bar{\mathbf{k}}} \rangle \right) + \{2nd\ term\}^\dagger$$

$$= \sum_{\bar{\mathbf{k}}} (\xi_{\bar{\mathbf{k}}\uparrow} + \xi_{-\bar{\mathbf{k}}\downarrow}) \left(u_{\bar{\mathbf{k}}}^* v_{\bar{\mathbf{k}}} \langle 0_{\bar{\mathbf{k}}} | 1_{\bar{\mathbf{k}}} \rangle + v_{\bar{\mathbf{k}}}^* v_{\bar{\mathbf{k}}} \langle 0_{\bar{\mathbf{k}}} | 0_{\bar{\mathbf{k}}} \rangle \right)$$

$$-\sum_{\bar{\mathbf{k}}} \Delta_{\bar{\mathbf{k}}} u_{\bar{\mathbf{k}}} \left(u_{\bar{\mathbf{k}}}^* \langle 0_{\bar{\mathbf{k}}} | c_{\bar{\mathbf{k}}\uparrow}^\dagger c_{-\bar{\mathbf{k}}\downarrow}^\dagger | 0_{\bar{\mathbf{k}}} \rangle + v_{\bar{\mathbf{k}}}^* \langle 0_{\bar{\mathbf{k}}} | 0_{\bar{\mathbf{k}}} \rangle \right) + \{2nd\ term\}^\dagger$$

$$= \sum_{\bar{\mathbf{k}}} (\xi_{\bar{\mathbf{k}}\uparrow} + \xi_{-\bar{\mathbf{k}}\downarrow}) \left(u_{\bar{\mathbf{k}}}^* v_{\bar{\mathbf{k}}} \cdot 0 + v_{\bar{\mathbf{k}}}^* v_{\bar{\mathbf{k}}} \cdot 1 \right)$$

$$-\sum_{\bar{\mathbf{k}}} \Delta_{\bar{\mathbf{k}}} u_{\bar{\mathbf{k}}} \left(u_{\bar{\mathbf{k}}}^* \cdot 0 + v_{\bar{\mathbf{k}}}^* \cdot 1 \right) + \{2nd\ term\}^\dagger$$

$$= \sum_{\bar{\mathbf{k}}} (\xi_{\bar{\mathbf{k}}\uparrow} + \xi_{-\bar{\mathbf{k}}\downarrow}) |v_{\mathbf{k}}|^2 - \sum_{\bar{\mathbf{k}}} \Delta_{\bar{\mathbf{k}}} u_{\bar{\mathbf{k}}} v_{\bar{\mathbf{k}}}^* + \{2nd\ term\}^\dagger$$

$$= \sum_{\bar{\mathbf{k}}} \left[(\xi_{\bar{\mathbf{k}}\uparrow} + \xi_{-\bar{\mathbf{k}}\downarrow}) |v_{\mathbf{k}}|^2 - \Delta_{\bar{\mathbf{k}}} u_{\bar{\mathbf{k}}} v_{\bar{\mathbf{k}}}^* - \Delta_{\bar{\mathbf{k}}}^* u_{\bar{\mathbf{k}}}^* v_{\bar{\mathbf{k}}} \right] \tag{41b}$$

5. MICROSCOPICALLY DETERMINED ENTROPY S

What is needed is a measure of the number of states in the system, or equivalently, the bookkeeping working with various types of states in the system. Consider the first measure, and refer to information theory and give a statistical definition as

$$S = k_B \ln \Omega \tag{42}$$

where Ω is the total number of system states. Now consider that a fundamental assumption of statistical thermodynamics or mechanics, is that the occupation of any microstate, labeled by index i, is assumed to be equally probable. That is, the probability p_i is given by

$$p_i = \frac{1}{\Omega} \tag{43}$$

Inverting this relationship, and inserting into (42) gives

$$S = \left(\frac{1}{\Omega}\sum_i\right) k_B \ln\left(\frac{1}{p_i}\right) = \sum_i k_B \frac{1}{\Omega} \ln\left(\frac{1}{p_i}\right) = -k_B \sum_i p_i \ln p_i \tag{44}$$

This last relationship constitutes the second measure, the one we are seeking and will employ.

We will treat two types of particles in the system, one particle being the electron at some momentum–spin $\mathbf{k}\sigma$ state, and the other its complement, or absence, perhaps calling it a hole. Finally, the particle statistics is governed by the Fermi–Dirac relationship, which enlisting the dispersion condition in (29) for $E_{\mathbf{k}}$ under the high symmetry case, is

$$f_k = \frac{1}{1 + e^{\beta E_{\mathbf{k}}}} \tag{45}$$

To utilize S given in (44), assign for the particle

$$i \rightarrow \mathbf{k}; \qquad p_i \rightarrow f_k \tag{46a}$$

and

$$i \rightarrow \mathbf{k}; \qquad \bar{p}_i \rightarrow 1 - f_k \tag{46b}$$

for its absence. Then the total entropy is

$$S = S_{particle} + S_{absence} = -k_B \sum_{\mathbf{k}\sigma} p_{\mathbf{k}\sigma} \ln p_{\mathbf{k}\sigma} - k_B \sum_{\mathbf{k}\sigma} \bar{p}_{\mathbf{k}\sigma} \ln \bar{p}_{\mathbf{k}\sigma} \tag{47}$$

for the quasi-particles with energy $E_{\mathbf{k}}$. Because the sum over spin may be performed,

$$\begin{aligned} S &= -2k_B \sum_{\mathbf{k}} p_{\mathbf{k}} \ln p_{\mathbf{k}} - 2k_B \sum_{\mathbf{k}} \bar{p}_{\mathbf{k}} \ln \bar{p}_{\mathbf{k}} \\ &= -2k_B \sum_{\mathbf{k}} f_{\mathbf{k}} \ln f_{\mathbf{k}} - 2k_B \sum_{\mathbf{k}} (1 - f_{\mathbf{k}}) \ln(1 - f_{\mathbf{k}}) \end{aligned} \tag{48}$$

6. EVALUATING THE MICROSCOPIC ENTROPY S

Using the last expression in (47), inserting the occupation probability (45), the following is obtained,

$$S = -2k_B \left\{ \sum_{\mathbf{k}} f_{\mathbf{k}} \ln\left(\frac{1}{1 + e^{\beta E_{\mathbf{k}}}}\right) + \sum_{\mathbf{k}} (1 - f_{\mathbf{k}}) \ln\left(\frac{e^{\beta E_{\mathbf{k}}}}{1 + e^{\beta E_{\mathbf{k}}}}\right) \right\}$$

$$= -2k_B\left\{\sum_{\mathbf{k}}\left[f_{\mathbf{k}}\ln\left(\frac{e^{-\beta E_{\mathbf{k}}}}{e^{-\beta E_{\mathbf{k}}}+1}\right)+(1-f_{\mathbf{k}})\ln\left(\frac{1}{e^{-\beta E_{\mathbf{k}}}+1}\right)\right]\right\}$$

$$= -2k_B\left\{\sum_{\mathbf{k}}\left[-f_{\mathbf{k}}\beta E_{\mathbf{k}}+(-f_{\mathbf{k}}-1+f_{\mathbf{k}})\ln\left(1+e^{-\beta E_{\mathbf{k}}}\right)\right]\right\}$$

$$= -2k_B\left\{-\int_{-\infty,\ both\ E_{\mathbf{k}}\ br}^{\infty} D(\xi_{\mathbf{k}}(\varepsilon_{\mathbf{k}}))\beta f_{\mathbf{k}}(\xi_{\mathbf{k}}(\varepsilon_{\mathbf{k}}))E_{\mathbf{k}}(\xi_{\mathbf{k}}(\varepsilon_{\mathbf{k}}))d\xi_{\mathbf{k}}\right.$$
$$\left.-\int_{-\infty,\ both\ E_{\mathbf{k}}\ br}^{\infty} D(\xi_{\mathbf{k}}(\varepsilon_{\mathbf{k}}))\ln\left(1+e^{-\beta E_{\mathbf{k}}}\right)d\xi_{\mathbf{k}}\right\} \quad (49)$$

where $D(\xi_{\mathbf{k}}(\varepsilon_{\mathbf{k}}))$ is the density of states per unit energy. Explicitly performing the integrals for the two $E_{\mathbf{k}}$ branches $E_{\mathbf{k}}^{+} = +E_{\mathbf{k}}$ and $E_{\mathbf{k}}^{-} = -E_{\mathbf{k}}$,

$$S = 2k_B\left[\int_{-\infty,\ E_{\mathbf{k}}=E_{\mathbf{k}}^{+}}^{\infty} D(\xi_{\mathbf{k}}(\varepsilon_{\mathbf{k}}))\beta f_{\mathbf{k}}(\xi_{\mathbf{k}}(\varepsilon_{\mathbf{k}}))\left[+E_{\mathbf{k}}(\xi_{\mathbf{k}}(\varepsilon_{\mathbf{k}}))\right]d\xi_{\mathbf{k}}\right.$$
$$+\int_{-\infty,\ E_{\mathbf{k}}=E_{\mathbf{k}}^{-}}^{\infty} D(\xi_{\mathbf{k}}(\varepsilon_{\mathbf{k}}))\beta f_{\mathbf{k}}(\xi_{\mathbf{k}}(\varepsilon_{\mathbf{k}}))\left[-E_{\mathbf{k}}(\xi_{\mathbf{k}}(\varepsilon_{\mathbf{k}}))\right]d\xi_{\mathbf{k}}$$
$$+\int_{-\infty,\ E_{\mathbf{k}}=E_{\mathbf{k}}^{+}}^{\infty} D(\xi_{\mathbf{k}}(\varepsilon_{\mathbf{k}}))\ln\left(1+e^{-\beta E_{\mathbf{k}}}\right)d\xi_{\mathbf{k}}$$
$$\left.+\int_{-\infty,\ E_{\mathbf{k}}=E_{\mathbf{k}}^{-}}^{\infty} D(\xi_{\mathbf{k}}(\varepsilon_{\mathbf{k}}))\ln\left(1+e^{-\beta E_{\mathbf{k}}}\right)d\xi_{\mathbf{k}}\right]$$

$$= 2k_B\left[\int_{-\infty}^{\infty} D(\xi_{\mathbf{k}}(\varepsilon_{\mathbf{k}}))\beta f_{\mathbf{k}}(\xi_{\mathbf{k}}(\varepsilon_{\mathbf{k}}))\left[+E_{\mathbf{k}}(\xi_{\mathbf{k}}(\varepsilon_{\mathbf{k}}))-E_{\mathbf{k}}(\xi_{\mathbf{k}}(\varepsilon_{\mathbf{k}}))\right]d\xi_{\mathbf{k}}\right.$$
$$+\int_{-\infty,\ E_{\mathbf{k}}=E_{\mathbf{k}}^{+}}^{\infty} D(\xi_{\mathbf{k}}(\varepsilon_{\mathbf{k}}))\ln\left(1+e^{-\beta E_{\mathbf{k}}}\right)d\xi_{\mathbf{k}}$$
$$\left.+\int_{-\infty,\ E_{\mathbf{k}}=E_{\mathbf{k}}^{-}}^{\infty} D(\xi_{\mathbf{k}}(\varepsilon_{\mathbf{k}}))\ln\left(1+e^{-\beta E_{\mathbf{k}}}\right)d\xi_{\mathbf{k}}\right]$$

$$= 2k_B\left[\int_{-\infty}^{\infty} D(\xi_{\mathbf{k}}(\varepsilon_{\mathbf{k}}))\beta f_{\mathbf{k}}(\xi_{\mathbf{k}}(\varepsilon_{\mathbf{k}}))\cdot 0\cdot d\xi_{\mathbf{k}}\right.$$
$$+2\int_{0}^{\infty} D(\xi_{\mathbf{k}}(\varepsilon_{\mathbf{k}}))\ln\left(1+e^{-\beta E_{\mathbf{k}}^{+}}\right)d\xi_{\mathbf{k}}$$
$$\left.+2\int_{0}^{\infty} D(\xi_{\mathbf{k}}(\varepsilon_{\mathbf{k}}))\ln\left(1+e^{-\beta E_{\mathbf{k}}^{-}}\right)d\xi_{\mathbf{k}}\right]$$

$$\approx 4k_B\left[D(\varepsilon_F)\int_{0}^{\infty}\ln\left(1+e^{-\beta E_{\mathbf{k}}}\right)d\xi_{\mathbf{k}}\right.$$
$$\left.+D(\varepsilon_F)\int_{0}^{\infty}\frac{D(\xi_{\mathbf{k}}(\varepsilon_{\mathbf{k}}))}{D(\varepsilon_F)}\ln\left(1+e^{\beta E_{\mathbf{k}}}\right)d\xi_{\mathbf{k}}\right]$$

$$\approx 4k_B\left[D(\varepsilon_F)\int_0^\infty f_{\mathbf{k}}(\beta E_{\mathbf{k}})\frac{\xi_{\mathbf{k}}^2}{E_{\mathbf{k}}}d\xi_{\mathbf{k}} + \big\langle D\big(\xi_{\mathbf{k}}(\varepsilon_{\mathbf{k}})\big)\big\rangle_{av}\int_0^{\xi_{max}}\ln\big(1+e^{\beta E_{\mathbf{k}}}\big)d\xi_{\mathbf{k}}\right]$$

$$\approx 4k_B\left[D(\varepsilon_F)\int_0^\infty f_{\mathbf{k}}(\beta E_{\mathbf{k}})\frac{\xi_{\mathbf{k}}^2}{E_{\mathbf{k}}}d\xi_{\mathbf{k}}\right.$$

$$\left.+\big\langle D\big(\xi_{\mathbf{k}}(\varepsilon_{\mathbf{k}})\big)\big\rangle_{av}\frac{\xi_{max}}{2}\left\{\ln\big(1+e^{\beta\Delta_{\mathbf{k}}}\big)+\beta E_k(\xi_{max})\right\}\right]$$

$$\approx 4k_B\left[D(\varepsilon_F)\int_0^\infty f_{\mathbf{k}}(\beta E_{\mathbf{k}})\frac{\xi_{\mathbf{k}}^2}{E_{\mathbf{k}}}d\xi_{\mathbf{k}} + \big\langle D\big(\xi_{\mathbf{k}}(\varepsilon_{\mathbf{k}})\big)\big\rangle_{av}\frac{\xi_{max}}{2}\{\beta\Delta_{\mathbf{k}}+\beta\xi_{max}\}\right]$$

$$\approx 4k_B\left[D(\varepsilon_F)\int_0^\infty f_{\mathbf{k}}(\beta E_{\mathbf{k}})\frac{\xi_{\mathbf{k}}^2}{E_{\mathbf{k}}}d\xi_{\mathbf{k}} + \big\langle D\big(\xi_{\mathbf{k}}(\varepsilon_{\mathbf{k}})\big)\big\rangle_{av}\frac{\xi_{max}}{2}\beta\xi_{max}\right]$$

$$\approx 4k_B\left[D(\varepsilon_F)\int_0^\infty f_{\mathbf{k}}(\beta E_{\mathbf{k}})\frac{\xi_{\mathbf{k}}^2}{E_{\mathbf{k}}}d\xi_{\mathbf{k}} + \frac{\langle D(\xi_{\mathbf{k}}(\varepsilon_{\mathbf{k}}))\rangle_{av}\hbar\omega_D}{2}\beta\hbar\omega_D\right] \quad (50)$$

Note that the integral property $\int_0^\infty \ln(1+e^{-x})dx = \pi^2/12$ from Dwight (1961) is used for the upper $E_{\mathbf{k}}^+$ branch integral, whereas the maximum ξ_{max} is estimated as the Debye energy $\hbar\omega_D$ for the lower branch integral. Clearly, (50) represents the superconducting state. The normal state can be found by taking the gap parameter to zero, or

$$S_n = \lim_{\Delta_{\mathbf{k}}\to 0} S = 4k_B\left[D(\varepsilon_F)\frac{\pi^2}{12\beta} + \frac{\langle D(\xi_{\mathbf{k}}(\varepsilon_{\mathbf{k}}))\rangle_{av}\hbar\omega_D}{2}\beta\hbar\omega_D\right] \quad (51)$$

by inspection of the last line of (50). The superconducting value, is of course, $S_s = S$.

7. EVALUATING THE AVERAGED ENERGY E

Recall from (40) that the general expression for averaged energy E is

$$E = \sum_{\mathbf{k}}\left[(\xi_{\mathbf{k}\uparrow}+\xi_{-\mathbf{k}\downarrow})|v_{\mathbf{k}}|^2 - \Delta_{\mathbf{k}}u_{\mathbf{k}}v_{\mathbf{k}}^* - \Delta_{\mathbf{k}}^*u_{\mathbf{k}}^*v_{\mathbf{k}}\right] \quad (52)$$

Using (30) with $\vartheta_u = \vartheta_v = 0$, $u_{\mathbf{k}}$ and $v_{\mathbf{k}}$ are chosen as real scalars. Similarly one can set $\vartheta_\Delta = 0$, making $\Delta_{\bar{\mathbf{k}}}$ also real. And using band structure symmetry, a simplified form of E is found

$$E = 2\sum_{\mathbf{k}}\left[\xi_{\mathbf{k}}v_{\mathbf{k}}^2 - \Delta_{\mathbf{k}}u_{\mathbf{k}}v_{\mathbf{k}}\right]$$

$$= 2\sum_{\mathbf{k}}\left[\xi_{\mathbf{k}}\frac{1}{2}\left(1-\frac{\xi_{\mathbf{k}}}{E_{\mathbf{k}}}\right) - \Delta_{\mathbf{k}}\frac{1}{\sqrt{2}}\left(1+\frac{\xi_{\mathbf{k}}}{E_{\mathbf{k}}}\right)^{1/2}\frac{1}{\sqrt{2}}\left(1-\frac{\xi_{\mathbf{k}}}{E_{\mathbf{k}}}\right)^{1/2}\right]$$

$$
\begin{aligned}
&= \sum_{\mathbf{k};\ E_{\mathbf{k}}\ branches} [\xi_{\mathbf{k}} - E_{\mathbf{k}}] \\
&\approx D(\varepsilon_F)\int_0^{\hbar\omega_D}\left(\xi_{\mathbf{k}} - E_{\mathbf{k}}^{+}\right)d\xi_{\mathbf{k}} + D(\varepsilon_F)\int_{-\hbar\omega_D}^{0}\left(\xi_{\mathbf{k}} - E_{\mathbf{k}}^{+}\right)d\xi_{\mathbf{k}} \\
&\quad + D(\varepsilon_F)\int_0^{\hbar\omega_D}\left(\xi_{\mathbf{k}} - E_{\mathbf{k}}^{-}\right)d\xi_{\mathbf{k}} - D(\varepsilon_F)\int_{-\hbar\omega_D}^{0}\left(\xi_{\mathbf{k}} - E_{\mathbf{k}}^{-}\right)d\xi_{\mathbf{k}} \\
&= D(\varepsilon_F)\int_0^{\hbar\omega_D}(\xi_{\mathbf{k}} - E_{\mathbf{k}})d\xi_{\mathbf{k}} - D(\varepsilon_F)\int_{-\hbar\omega_D}^{0}(\xi_{\mathbf{k}} + E_{\mathbf{k}})d\xi_{\mathbf{k}} \\
&\quad + \left[D(\varepsilon_F)\int_0^{\hbar\omega_D}(\xi_{\mathbf{k}} + E_{\mathbf{k}})d\xi_{\mathbf{k}} + D(\varepsilon_F)\int_{-\hbar\omega_D}^{0}(\xi_{\mathbf{k}} - E_{\mathbf{k}})d\xi_{\mathbf{k}}\right] \\
&= D(\varepsilon_F)\int_0^{\hbar\omega_D}(\xi_{\mathbf{k}} - E_{\mathbf{k}})d\xi_{\mathbf{k}} - D(\varepsilon_F)\int_{-\hbar\omega_D}^{0}(\xi_{\mathbf{k}} + E_{\mathbf{k}})d\xi_{\mathbf{k}} \\
&= D(\varepsilon_F)\int_0^{\hbar\omega_D}(\xi_{\mathbf{k}} - E_{\mathbf{k}})d\xi_{\mathbf{k}} + D(\varepsilon_F)\int_{\hbar\omega_D}^{0}(-\bar{\xi}_{\mathbf{k}} + E_{\mathbf{k}})d\bar{\xi}_{\mathbf{k}} \\
&= 2D(\varepsilon_F)\int_0^{\hbar\omega_D}(\xi_{\mathbf{k}} - E_{\mathbf{k}})d\xi_{\mathbf{k}} \\
&= 2D(\varepsilon_F)\left[\int_0^{\hbar\omega_D}\xi_{\mathbf{k}}d\xi_{\mathbf{k}} - \int_0^{\hbar\omega_D}E_{\mathbf{k}}d\xi_{\mathbf{k}}\right] \\
&= 2D(\varepsilon_F)\left[\frac{(\hbar\omega_D)^2}{2} - \int_0^{\hbar\omega_D}\sqrt{\xi_{\mathbf{k}}^2 + \Delta_{\mathbf{k}}^2}\,d\xi_{\mathbf{k}}\right] \\
&= D(\varepsilon_F)(\hbar\omega_D)^2\left\{1 - \sqrt{1 + \left(\frac{\Delta_{\mathbf{k}}}{\hbar\omega_D}\right)^2}\right. \\
&\quad \left. - \left(\frac{\Delta_{\mathbf{k}}}{\hbar\omega_D}\right)^2 \ln\left[\frac{\hbar\omega_D}{\Delta_{\mathbf{k}}} + \sqrt{1 + \left(\frac{\hbar\omega_D}{\Delta_{\mathbf{k}}}\right)^2}\right]\right\}
\end{aligned}
\tag{53}
$$

with indefinite integral evaluation employed (Dwight, 1961). Clearly, (53) represents the superconducting state. The normal state can be found by taking the gap parameter to zero, or

$$E_n = \lim_{\Delta_{\mathbf{k}} \to 0} E = 0 \tag{54}$$

by inspection of the last line of (53). The superconducting value, is of course, $E_s = E$.

8. CRITICAL MAGNETIC FIELD FROM MICROSCOPIC APPROACH

From the energy E [(53) and (54)] and entropy S [(50) and (51)] expressions, the final Helmholtz free energy F can be written compactly for both the normal and superconducting conditions,

$$F = D(\varepsilon_F)(\hbar\omega_D)^2 \left\{ \begin{array}{l} 0, \quad normal \\ 1 - \sqrt{1 + \left(\frac{\Delta_{\mathbf{k}}}{\hbar\omega_D}\right)^2} - \left(\frac{\Delta_{\mathbf{k}}}{\hbar\omega_D}\right)^2 \ln\left[\frac{\hbar\omega_D}{\Delta_{\mathbf{k}}} + \sqrt{1 + \left(\frac{\hbar\omega_D}{\Delta_{\mathbf{k}}}\right)^2}\right], \\ \quad supercond \end{array} \right\}$$
$$- D(\varepsilon_F) \left\{ \begin{array}{ll} \frac{\pi^2}{3}(k_B T)^2, & n \\ 4\int_0^\infty f_{\mathbf{k}}(\beta E_{\mathbf{k}}) \frac{\xi_{\mathbf{k}}^2}{E_{\mathbf{k}}} d\xi_{\mathbf{k}}, & sc \end{array} \right\}$$
$$+ 2\langle D(\xi_{\mathbf{k}}(\varepsilon_{\mathbf{k}}))\rangle_{av}(\hbar\omega_D)^2 \tag{55}$$

By Kittel (1968), the thermodynamic physics allows one to express the difference of the free energies in the normal and superconducting states, at zero applied magnetic field $H_a = 0$, as related to the square of the critical magnetic field H_c:

$$F_n(T,0) - F_s(T,0) = \frac{1}{8\pi^2}(H_c)^2 \tag{56}$$

From (55), we can write down the free energies in the two states required in (56),

$$F_n = D(\varepsilon_F)(\hbar\omega_D)^2 \cdot 0 - D(\varepsilon_F)\frac{\pi^2}{3}(k_B T)^2 + 2\langle D(\xi_{\mathbf{k}}(\varepsilon_{\mathbf{k}}))\rangle_{av}(\hbar\omega_D)^2$$
$$= -D(\varepsilon_F)\frac{\pi^2}{3}(k_B T)^2 + 2\langle D(\xi_{\mathbf{k}}(\varepsilon_{\mathbf{k}}))\rangle_{av}(\hbar\omega_D)^2 \tag{57a}$$

$$F_s = D(\varepsilon_F)(\hbar\omega_D)^2 \left\{ 1 - \sqrt{1 + \left(\frac{\Delta_{\mathbf{k}}}{\hbar\omega_D}\right)^2} - \left(\frac{\Delta_{\mathbf{k}}}{\hbar\omega_D}\right)^2 \ln\left[\frac{\hbar\omega_D}{\Delta_{\mathbf{k}}} + \sqrt{1 + \left(\frac{\hbar\omega_D}{\Delta_{\mathbf{k}}}\right)^2}\right]\right\}$$
$$- 4D(\varepsilon_F)\int_0^\infty f_{\mathbf{k}}(\beta E_{\mathbf{k}})\frac{\xi_{\mathbf{k}}^2}{E_{\mathbf{k}}} d\xi + 2\langle D(\xi_{\mathbf{k}}(\varepsilon_{\mathbf{k}}))\rangle_{av}(\hbar\omega_D)^2 \tag{57b}$$

Taking the difference of (57) equations, the square of the critical magnetic field H_c becomes using (56)

$$\frac{1}{8\pi}(H_c(T))^2 = D(\varepsilon_F)(\hbar\omega_D)^2\left\{\sqrt{1+\left(\frac{\Delta_{\mathbf{k}}}{\hbar\omega_D}\right)^2} - 1 + \left(\frac{\Delta_{\mathbf{k}}}{\hbar\omega_D}\right)^2 \ln\left[\frac{\hbar\omega_D}{\Delta_{\mathbf{k}}} + \sqrt{1+\left(\frac{\hbar\omega_D}{\Delta_{\mathbf{k}}}\right)^2}\right]\right\} - D(\varepsilon_F)\frac{\pi^2}{3}(k_BT)^2\left\{1 - \frac{12\beta^2}{\pi^2}\int_0^\infty f_{\mathbf{k}}(\beta E_{\mathbf{k}})\frac{\xi_{\mathbf{k}}^2}{E_{\mathbf{k}}}d\xi_{\mathbf{k}}\right\} \quad (58)$$

Critical magnetic field H_{c0} at $T = 0$ may be determined from (58), realizing that the quadratic term in T as well as the integral term containing $f_{\mathbf{k}}(\beta E_{\mathbf{k}})$ vanishes because of the limiting ratio $\Delta_{\mathbf{k}}/k_BT$,

$$\begin{aligned}\lim_{T\to 0}\left[\frac{1}{8\pi}(H_c(T))^2\right] &\approx D(\varepsilon_F)(\hbar\omega_D)^2\left\{1 + \frac{1}{2}\left(\frac{\Delta_{\mathbf{k}}}{\hbar\omega_D}\right)^2 - 1 + \left(\frac{\Delta_{\mathbf{k}}}{\hbar\omega_D}\right)^2 \ln\left[\frac{\hbar\omega_D}{\Delta_{\mathbf{k}}} + \frac{\hbar\omega_D}{\Delta_{\mathbf{k}}}\right]\right\} \\ &\quad - D(\varepsilon_F)\frac{\pi^2}{3}\lim_{T\to 0}(k_BT)^2 - 4D(\varepsilon_F)\lim_{T\to 0}\left[\int_0^\infty f_{\mathbf{k}}(\beta E_{\mathbf{k}})\frac{\xi_{\mathbf{k}}^2}{E_{\mathbf{k}}}d\xi_{\mathbf{k}}\right] \\ &= D(\varepsilon_F)\left[\frac{1}{2}(\Delta_{\mathbf{k}}(0))^2 + (\Delta_{\mathbf{k}}(0))^2\ln\left\{2\frac{\hbar\omega_D}{\Delta_{\mathbf{k}}}\right\}\right] \\ &= D(\varepsilon_F)\left[(\Delta_{\mathbf{k}}(0))^2\left\{\frac{1}{2} + \ln 2\right\} - (\Delta_{\mathbf{k}}(0))^2\ln\left\{\frac{\Delta_{\mathbf{k}}}{\hbar\omega_D}\right\}\right] \\ &= D(\varepsilon_F)\left[(\Delta_{\mathbf{k}}(0))^2\left\{\frac{1}{2} + \ln 2\right\} - (\hbar\omega_D)^2\frac{\Delta_{\mathbf{k}}(0)}{\hbar\omega_D}\ln\left\{\left(\frac{\Delta_{\mathbf{k}}}{\hbar\omega_D}\right)^{\frac{\Delta_{\mathbf{k}}}{\hbar\omega_D}}\right\}\right] \\ &= D(\varepsilon_F)(\Delta_{\mathbf{k}}(0))^2\left\{\frac{1}{2} + \ln 2\right\}\end{aligned} \quad (59)$$

Here the fact that $\Delta_{\mathbf{k}}/\hbar\omega_D \to 0$ can be treated as vanishing small, allows the logarithmic form $\ln(x^x)$ in (59) to be dropped, $\lim_{x\to 0}\ln(x^x) \to 0$. Thus,

$$H_{c0} = \sqrt{4\pi D(\varepsilon_F)\{1 + 2\ln 2\}}\Delta_{\mathbf{k}}(0) \quad (60)$$

9. SPECIFIC HEAT OF ELECTRONS FOR THE SYSTEM

The electronic system specific heat or heat capacity in the normal state can be found from heat exchanged Q_{el} when it is energy $E_{el}^{heat,\,finite\ T}$ from thermodynamic considerations (Reif, 1965; Kittel, 1968),

$$C_{el,n}(T) = \frac{\partial Q_{el}}{\partial T} = \frac{\partial E_{el}^{heat,\,finite\ T}}{\partial T} = D_{both\ spin\ types}(\varepsilon_F) \int_0^\infty \frac{\partial f(\varepsilon, T)}{\partial T} (\varepsilon - \varepsilon_F) d\varepsilon \tag{61}$$

For example, in 3D parabolic band structure systems

$$C_{el,n}^{3D}(T) = \gamma_{3D} T; \qquad \gamma_{3D} = \frac{2}{3}\pi^2 D_{per\ spin}(\varepsilon_F)(k_B)^2 \tag{62}$$

For the system in a superconducting state, an alternative thermodynamic physical approach can be utilized, giving for heat capacity (Reif, 1965)

$$C_{el,s}(T) = T\left(\frac{\partial S_s}{\partial T}\right)_{\substack{const\ N_{el}\\ or\ vol}} = -\beta\left(\frac{\partial S_s}{\partial \beta}\right)_{\substack{const\ N_{el}\\ or\ vol}} \tag{63}$$

using the thermodynamic partial derivative. Refer to (50) to obtain S_s, using an intermediate expression with explicit logarithmic form for the first integral,

$$\begin{aligned} C_{el,s}(T) &= -\beta \frac{\partial}{\partial \beta}\left\{ 4k_B D(\varepsilon_F) \int_0^\infty \ln\left(1 + e^{-\beta E_{\mathbf{k}}}\right) d\xi_{\mathbf{k}} \right. \\ &\quad \left. + 2k_B \left\langle D\left(\xi_{\mathbf{k}}(\varepsilon_{\mathbf{k}})\right)\right\rangle_{av} (\hbar\omega_D)^2 \beta \right\} \\ &= 4k_B \beta D(\varepsilon_F) \int_0^\infty f(\beta E_{\mathbf{k}}) \left(E_{\mathbf{k}} + \beta \frac{\partial E_{\mathbf{k}}}{\partial \beta} \right) d\xi_{\mathbf{k}} \\ &\quad - 2\beta k_B \left\langle D\left(\xi_{\mathbf{k}}(\varepsilon_{\mathbf{k}})\right)\right\rangle_{av} (\hbar\omega_D)^2 \\ &= 4k_B \beta D(\varepsilon_F) \int_0^\infty f(\beta E_{\mathbf{k}}) \left(E_{\mathbf{k}} + \frac{\beta}{2E_{\mathbf{k}}} \frac{\partial (\Delta_{\mathbf{k}}(\beta))^2}{\partial \beta} \right) d\xi_{\mathbf{k}} \\ &\quad - 2\beta k_B \left\langle D\left(\xi_{\mathbf{k}}(\varepsilon_{\mathbf{k}})\right)\right\rangle_{av} (\hbar\omega_D)^2 \end{aligned} \tag{64}$$

Eq. (64) represents the superconducting state. The normal state can be found by taking the gap parameter to zero, or by inspection of the last line of (64), or taking the limit as $T \to T_c^+$ on the upper side of the critical temperature,

$$C_{el,n}(T) = \lim_{\Delta_{\mathbf{k}}\to 0} C_{el,s}(T)$$
$$= 4k_B\beta D(\varepsilon_F)\int_0^{\infty} f(\beta\xi_{\mathbf{k}})\xi_{\mathbf{k}}d\xi_{\mathbf{k}} - 2\beta k_B\langle D(\xi_{\mathbf{k}}(\varepsilon_{\mathbf{k}}))\rangle_{av}(\hbar\omega_D)^2 \quad (65)$$

10. SLOPE OF THE CRITICAL MAGNETIC FIELD AT THE CRITICAL TEMPERATURE

Thermodynamically using statistical physics, it is known that the difference of the heat capacities at zero applied magnetic field, is related to the square of the derivative of the magnetic field with respect to the temperature T at the critical temperature T_c (Kittel, 1968),

$$C_{el,n}(T_c, 0) - C_{el,s}(T_c, 0) = \frac{T_c}{4\pi}\left(\left.\frac{dH_c(T)}{dT}\right|_{T_c}\right)^2 \quad (66)$$

The left hand side of (66) can be evaluated by approaching T_c from above and below T_c for, respectively, the normal and superconducting heat capacities. The heat capacity difference is then from (64) and (65),

$$C_{el,n}\left(T_c^+, 0\right) - C_{el,s}\left(T_c^-, 0\right)$$
$$= 4k_B\beta_c D(\varepsilon_F)\left[\int_0^{\infty} f(\beta E_{\mathbf{k}})\left(E_{\mathbf{k}} - \xi_{\mathbf{k}} + \frac{\beta}{2E_{\mathbf{k}}}\frac{\partial(\Delta_{\mathbf{k}}(\beta))^2}{\partial\beta}\right)d\xi_{\mathbf{k}}\right]\Bigg|_{T\to T_c^-}$$
$$= 2k_B\beta_c D(\varepsilon_F)\left[\int_0^{\infty} f(\beta E_{\mathbf{k}})\frac{\beta}{E_{\mathbf{k}}}\frac{\partial(\Delta_{\mathbf{k}}(\beta))^2}{\partial\beta}d\xi_{\mathbf{k}}\right]\Bigg|_{T\to T_c^-}$$
$$\approx 2k_B(\beta_c)^2 D(\varepsilon_F)\left.\frac{\partial(\Delta_{\mathbf{k}}(\beta))^2}{\partial\beta}\right|_{\beta_c}\left[\int_0^{\infty} f(\beta E_{\mathbf{k}})\frac{1}{\Delta_{\mathbf{k}}(0)/2}d\xi_{\mathbf{k}}\right]\Bigg|_{T\to T_c^-}$$
$$= 4k_B\frac{(\beta_c)^2}{\Delta_{\mathbf{k}}(0)}D(\varepsilon_F)\left.\frac{\partial(\Delta_{\mathbf{k}}(\beta))^2}{\partial\beta}\right|_{\beta_c}\frac{1}{\beta_c}F_0(\xi_F = 0)$$
$$= 4k_B\frac{\beta_c}{\Delta_{\mathbf{k}}(0)}D(\varepsilon_F)\left.\frac{\partial(\Delta_{\mathbf{k}}(\beta))^2}{\partial\beta}\right|_{\beta_c}\ln 2 \quad (67)$$

The double $E_{\mathbf{k}}$ limit as $\xi_{\mathbf{k}}$ and $\Delta_{\mathbf{k}}(\beta)$ approach zero in part of the integral argument range, is dealt with by using an average of the gap parameter. Complete Fermi–Dirac integrals, $F_j(x) = (1/\Gamma(j+1))\int_0^{\infty}\frac{t^j}{e^{t-x}+1}dt$. For $j = 0$, $F_0(x) = \int_0^{\infty}\frac{dt}{e^{t-x}+1}$, $\Gamma(1) = 1$; $F_0(x) = \ln(1 + e^x)$.

Equating (67) with (66) yields the solution to critical magnetic field slope at T_c,

$$\left.\frac{dH_c(T)}{dT}\right|_{T_c} = \sqrt{\frac{4\pi k_B \beta_c}{T_c}\frac{4\ln 2}{\Delta_{\mathbf{k}}(0)}D(\varepsilon_F)\left.\frac{\partial(\Delta_{\mathbf{k}}(\beta))^2}{\partial\beta}\right|_{\beta_c}}$$
$$= \sqrt{16\pi (k_B)^2(\beta_c)^2\frac{\ln 2}{\Delta_{\mathbf{k}}(0)}D(\varepsilon_F)\left.\frac{\partial(\Delta_{\mathbf{k}}(\beta))^2}{\partial\beta}\right|_{\beta_c}} \quad (68)$$

11. RELATING THE DISORDER POTENTIAL ENERGY TO THE GAP PARAMETER

The gap parameter used in the previous sections is not the regular or ordinary gap parameter. Rather, it has been modified by the impurity scattering potential energy, and so that altered gap is derived here. There are less typical or expected quantum many-body Green's functions, often referred to as anomalous Green's functions. One of them which will be of use is defined as (Bruus & Flensberg, 2004)

$$F_{\mathbf{k}\downarrow\uparrow}(\mathbf{k},\tau) = -\left\langle T_\tau\left\{c^\dagger_{-\mathbf{k}\downarrow}(\tau)c^\dagger_{\mathbf{k}\uparrow}(0)\right\}\right\rangle \quad (69)$$

T_τ is the time ordering operator in the imaginary Matsubara time τ frame. The bracketed operation indicates an ensemble statistical average over any complete eigenstate set ν. Thus for some operator O,

$$\langle O\rangle = \frac{1}{Z}\sum_\nu \langle\nu|O|\nu\rangle e^{-\beta E_\nu}; \qquad Z = \sum_\nu e^{-\beta E_\nu} \quad (70)$$

Here Z is the partition function, formed by the exponential sum using the system Hamiltonian eigenenergies E_ν. The signature property of the anomalous Green's function is that it does not have mixed raising and lowering operators as the ordinary Green's function (Fetter & Walecka, 1971)

$$G_{\uparrow\uparrow}(\mathbf{k},\tau) = -\left\langle T_\tau\left\{c_{\mathbf{k}\uparrow}(\tau)c^\dagger_{\mathbf{k}\uparrow}(0)\right\}\right\rangle \quad (71)$$

These two types of Green's functions can be related by equations of motion,

$$\partial_\tau G_{\uparrow\uparrow}(\mathbf{k},\tau) = -\delta(\tau) - \xi_{\mathbf{k}}G_{\uparrow\uparrow}(\mathbf{k},\tau) + \Delta_{\mathbf{k}}F_{\downarrow\uparrow}(\mathbf{k},\tau) \quad (72a)$$
$$\partial_\tau F_{\downarrow\uparrow}(\mathbf{k},\tau) = \qquad -\xi_{\mathbf{k}}F_{\downarrow\uparrow}(\mathbf{k},\tau) + \Delta_{\mathbf{k}}G_{\uparrow\uparrow}(\mathbf{k},\tau) \quad (72b)$$

It is worth noting that the 2nd quantized raising and lowering electron operators employed in (69) and (71) are perturbed many-body operators.

Eq. (36) relates the total potential energy to the gap parameter using 2nd quantized electron operators. From it we need to find actual relations which explicitly demonstrate the relationship between the disorder effect via the impurity scattering potential energy, to the gap parameter. The starting point is recognizing that the expression uses the anomalous Green's function, converting into the Matsubara frequency domain, retrieving the 4-momentum space formula for it, eventually performing an integration.

$$\begin{aligned}
\Delta_{\mathbf{k}} &= \sum_{\mathbf{k}'}^{|\xi_{\mathbf{k}'}|<\omega_D} V'_{\mathbf{kk}'}\langle c_{-\mathbf{k}'\downarrow} c_{\mathbf{k}\uparrow}\rangle \\
&\approx V'^{\,RPA}_{eff,av} \sum_{\mathbf{k}'}^{|\xi_{\mathbf{k}'}|<\omega_D} F^{*}_{\downarrow\uparrow}(\mathbf{k},\tau)|_{\tau\to 0} \\
&= V'^{\,RPA}_{eff,av} \sum_{\mathbf{k}}^{|\xi_{\mathbf{k}'}|<\omega_D} \sum_{ik_n} e^{-ik_n\tau} F^{*}_{\downarrow\uparrow}(\mathbf{k}, ik_n)|_{\tau\to 0} \\
&= V'^{\,RPA}_{eff,av} \sum_{\mathbf{k}}^{|\xi_{\mathbf{k}'}|<\omega_D} \sum_{ik_n} e^{-ik_n\cdot 0^+} F^{*}_{\downarrow\uparrow}(\mathbf{k}, ik_n) \\
&= V'^{\,RPA}_{eff,av} \sum_{\mathbf{k}}^{|\xi_{\mathbf{k}'}|<\omega_D} \sum_{ik_n} e^{-ik_n\cdot 0^+} \left[\frac{-\Delta^{*}_{\mathbf{k}}}{(ik_n)^2-(E_{\mathbf{k}})^2}\right]^{*} \\
&= -V'^{\,RPA}_{eff,av} \sum_{\mathbf{k}}^{|\xi_{\mathbf{k}'}|<\omega_D} \Delta_{\mathbf{k}} \frac{1}{\beta} \sum_{ik_n} e^{-ik_n\cdot 0^+} \frac{1}{(ik_n)^2-(E_{\mathbf{k}})^2} \\
&= -V'^{\,RPA}_{eff,av} \sum_{\mathbf{k}}^{|\xi_{\mathbf{k}'}|<\omega_D} \Delta_{\mathbf{k}} \frac{1}{\beta} \sum_{ik_n} e^{-ik_n\cdot 0^+} \frac{1}{(ik_n-|E_{\mathbf{k}}|)(ik_n+|E_{\mathbf{k}}|)}
\end{aligned} \tag{73}$$

$\Delta_{\mathbf{k}}$ is readily evaluated by using the contour integration property for the complex valued function $g_0(z)$ with poles in the complex z-plane,

$$\frac{1}{\beta}\sum_{ik_n} g_0(ik_n)e^{ik_n\tau}\bigg|_{\tau>0} = \sum_{j=1}^{M} f(z_j)e^{z_j\tau}\operatorname*{Res}_{z=z_j}\left[g_0(z)\right] \tag{74}$$

where f is just the Fermi–Dirac function. Noting in (72) we used the following for $F_{\downarrow\uparrow}(\mathbf{k},\tau)$ and identifying $g_0(z)$ as

$$F_{\downarrow\uparrow} = \frac{-\Delta_{\mathbf{k}}^*}{(ik_n)^2 - (E_{\mathbf{k}})^2}; \qquad g_0(z) = \frac{1}{(z - |E_{\mathbf{k}}|)(z + |E_{\mathbf{k}}|)} \tag{75}$$

one finds

$$\begin{aligned}
\Delta_{\mathbf{k}} &= -V'^{RPA}_{eff,av} \sum_{\mathbf{k}}^{|\xi_{\mathbf{k}'}|<\omega_D} \Delta_{\mathbf{k}} \frac{f(\beta E^+_{\mathbf{k}}) - f(\beta E^-_{\mathbf{k}})}{2E^+_{\mathbf{k}}} \\
&= -V'^{RPA}_{eff,av} \sum_{\mathbf{k}}^{|\xi_{\mathbf{k}'}|<\omega_D} \Delta_{\mathbf{k}} \frac{f(\beta|E_{\mathbf{k}}|) - f(-\beta|E_{\mathbf{k}}|)}{2|E_{\mathbf{k}}|} \\
&= -V'^{RPA}_{eff,av} \sum_{\mathbf{k}}^{|\xi_{\mathbf{k}'}|<\omega_D} \Delta_{\mathbf{k}} \frac{f(\beta E_{\mathbf{k}}) - f(-\beta E_{\mathbf{k}})}{2E_{\mathbf{k}}} \\
&= V'^{RPA}_{eff,av} \sum_{\mathbf{k}}^{|\xi_{\mathbf{k}'}|<\omega_D} \Delta_{\mathbf{k}} \frac{1 - 2f(\beta E_{\mathbf{k}})}{2E_{\mathbf{k}}}
\end{aligned} \tag{76}$$

where we dropped the magnitude signs on the dispersion relation, taking by default the positive branch.

Considering the simplest condition where the gap parameter becomes momentum independent, $\Delta_{\mathbf{k}} \to \Delta$, will factor out of the summation, leaving it to be divided out,

$$\begin{aligned}
1 &= V'^{RPA}_{eff,av} \sum_{\mathbf{k}}^{|\xi_{\mathbf{k}'}|<\omega_D} \frac{1 - 2f(\beta E_{\mathbf{k}})}{2E_{\mathbf{k}}} \\
&= V'^{RPA}_{eff,av} \int_{-\varepsilon_F}^{\infty} D(\xi_{\mathbf{k}}) \frac{1 - 2f(\beta E_{\mathbf{k}})}{2E_{\mathbf{k}}} d\xi_{\mathbf{k}} \\
&\approx V'^{RPA}_{eff,av} D(\varepsilon_F) \int_{-\varepsilon_F}^{\infty} \frac{1 - 2f(\beta E_{\mathbf{k}})}{2E_{\mathbf{k}}} d\xi_{\mathbf{k}} \\
&\approx V'^{RPA}_{eff,av} D(\varepsilon_F) \int_{-\omega_D}^{\omega_D} \frac{1 - 2f(\beta E_{\mathbf{k}})}{2E_{\mathbf{k}}} d\xi_{\mathbf{k}} \\
&\approx V'^{RPA}_{eff,av} D(\varepsilon_F) \int_{-\omega_D}^{\omega_D} \frac{\tanh\left(\frac{\beta}{2}\sqrt{(\xi_{\mathbf{k}})^2 + |\Delta|^2}\right)}{2\sqrt{(\xi_{\mathbf{k}})^2 + |\Delta|^2}} d\xi_{\mathbf{k}}
\end{aligned} \tag{77}$$

The final equality in (77) provides the relationship between the average effective RPA potential energy and the gap parameter. [Note that $d(\varepsilon_F) = d_{st}(\varepsilon_F) = D(\varepsilon_F)/\boldsymbol{\mathcal{V}}$, where $\boldsymbol{\mathcal{V}}$ is the volume in 3D, which is replaced by area A_{2D} in 2D.]

A non-transcendental relationship is found as $T \to 0$, for the superconducting gap parameter, when the Coulomb-phonon and impurity potential

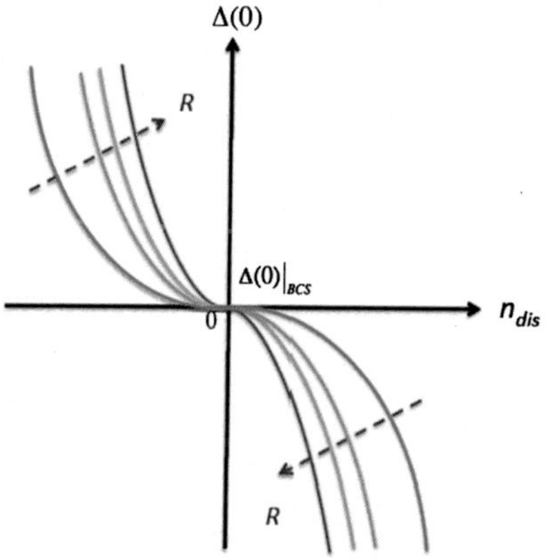

Figure 1 Variation of the superconducting gap with the disorder density. Parameterization in terms of potential energy ratio strength *R*.

energy ratio $R^{V}_{s,imp;CP}$ is small, $V'^{RPA}_{imp,av}/V'^{RPA,CP}_{eff,av} = n_{imp}R^{V}_{s,imp;CP}$,

$$\Delta(0)=\Delta(0)|_{BCS}\left(\Delta(0)\big|_{\substack{BCS\\ \exp}}\right)^{-n_{dis}R};\qquad \Delta(0)|_{BCS}=2\hbar\omega_D\Delta(0)\big|_{\substack{BCS\\ \exp}};$$

$$\Delta(0)\big|_{\substack{BCS\\ \exp}}=e^{-\frac{2}{V'^{RPA,CP}_{eff,av}}\frac{1}{d(\varepsilon_F)}} \tag{78}$$

where

$$\left|\Delta(0)\right|_{\substack{BCS\\ pre}}=2\hbar\omega_D;\qquad \left|\Delta(0)\right|_{\substack{BCS\\ \exp}}=e^{-\frac{1}{V'^{RPA,CP}_{eff,av}D(\varepsilon_F)}};$$

$$\gamma_{rp}=\frac{1}{1+n_{imp}R^{V}_{s,imp;CP}} \tag{79}$$

Here BCS as a subscript serves to identify the BCS like behavior in the prefactor coefficient and the exponential terms, whereas γ_{rp} is the exponential modification power which acts on the entire third quantity in (78), unless is close to unity and simplifies. What may occur is that increasing the number of impurities may reduce the disorder in the material when constructively affecting the arrangement of the atomic lattice, so the actual disorder number, n_{dis} will no longer equal n_{imp}, and could go negative.

The result in (78) is sketched in Fig. 1, where R stands for magnitude $|R^{V}_{s,imp;CP}|$, and $\Delta(0)$ is plotted against n_{dis}. Notice, negative going values of n_{dis} increase $\Delta(0)$ whereas positive going values decrease $\Delta(0)$. Parametrization in terms of R is done.

12. CONCLUSIONS

Disorder as it microscopically affects superconducting properties, has been delineated in the treatment given here. Useful, compact analytical expressions for many parameters of interest, including critical magnetic field H_c, critical magnetic field slope dH_c/dT, and heat capacity, and superconducting gap $\Delta(T)$, are derived. There is also an implicit relationship between critical temperature T_c and gap parameter $\Delta(T)$ because of Δ's dependence on the same disorder potential energy quantity as T_c. The treatment shown here opens up the possibility that disorder could possibly increase $\Delta(T)$ as well as T_c, something not seen in the 1980s, but observed in some circumstances as mentioned previously in the Introduction.

The results are not inconsistent with the original BCS theory (Bardeen, Cooper, & Schrieffer, 1957), although formulas provided here are derived in a more streamlined fashion, not necessarily relying on bandstructure symmetries originally utilized in either reciprocal $\mathbf{k}$-space or spin index σ.

ACKNOWLEDGMENTS

This contribution arose while working on the project Lower Dimensional Materials for Naval Applications, involving our Electronics Science & Technology, Material Science & Technology, and Chemistry Divisions, with the Plasma Physics Division also participating as a collaborating division, at the Naval Research Laboratory, Washington, D.C. Interactions with all of the project researchers in these divisions, has informed the contents of the present work, over the course of the project, from October 2013 to October 2017. I mention with particular gratitude the many interesting discussions with Dr. Michael S. Osofsky of MSTD, on electron–electron interactions, superconductivity, metal–insulator transition, and many other topics related to 2D solid state material systems.

REFERENCES

Abrahams, E., Anderson, P. W., Licciardello, D. C., & Ramakrishnan, T. V. (1979). Scaling theory of localization: Absence of quantum diffusion in two dimensions. *Physical Review Letters*, *42*, 673–676.

Abrikosov, A. A. (2004). Nobel lecture: Type-II superconductors and the vortex lattice. *Reviews of Modern Physics*, *76*, 975.

Abrikosov, A. A., & Gor'kov, L. P. (1959a). Superconducting alloys at finite temperatures. *Soviet Physics, JETP*, *9*, 220 (in Russian, *Journal of Experimental and Theoretical Physics (USSR)*, *36*, 319, Jan. 1959).

Abrikosov, A. A., & Gor'kov, L. P. (1959b). On the theory of superconducting alloys 1. The electrodynamics of alloys at absolute zero. *Soviet Physics, JETP*, *35*, 1090 (in Russian, *Journal of Experimental and Theoretical Physics (USSR)*, *35*, 1558, Dec. 1958).

Abrikosov, A. A., & Gor'kov, L. P. (1961). Contributions to the theory of superconducting alloys with paramagnetic impurities. *Soviet Physics, JETP, 12*, 1243 (in Russian, *Journal of Experimental and Theoretical Physics (USSR), 39*, 1781, Dec. 1960).

Abrikosov, A. A., Gorkov, L. P., & Dzyaloshinski, I. E. (1963). *Methods of quantum field theory in statistical physics* (R. A. Silverman, Trans. Ed.). Englewood Cliffs, NJ: Prentice-Hall, Inc.

Abrikosov, A. A., Gorkov, L. P., & Dzyaloshinski, I. Ye. (1965). *Quantum field theoretical methods in statistical physics* (D. E. Brown, Trans., D. ter Haar, Ed.). Oxford: Pergamon Press Ltd. (Original ed. publ. Fizmatgiz, Moscow, 1962 & added mater. 1964).

Anderson, P. W. (1959). Theory of dirty superconductors. *Journal of Physics and Chemistry of Solids, 11*, 26–30.

Attanasi, A. (2008). *Competition between superconductivity and charge density waves: The role of disorder* (Ph.D. thesis). Sapienza Universita di Roma.

Balatsky, A. V., Vekhter, I., & Zhu, J.-X. (2006). Impurity-induced states in conventional and unconventional superconductors. *Reviews of Modern Physics, 78*, 373.

Bardeen, J., Cooper, L. N., & Schrieffer, J. R. (1957). Theory of superconductivity. *Physical Review, 108*, 1175.

Bartolf, H. (2016). *Fluctuation mechanisms in superconductors*. Springer Spektrum.

Bellafi, B., Haddad, S., & Charfi-Kaddour, S. (2009). Disorder-induced superconductivity in ropes of carbon nanotubes. *Physical Review B, 80*, 075401.

Bonetto, C., Israeloff, N. E., Pokrovskiy, N., & Bojko, R. (1998). Field induced superconductivity in disordered wire networks. *Physical Review, 58*, 128.

Bruus, H., & Flensberg, K. (2004). *Many-body quantum theory in condensed matter physics – An introduction*. Oxford University Press (reprinted in 2013).

Bugoslavsky, Y., Cohen, L. F., Perkins, G. K., Polichetti, M., Tate, T. J., Gwilliam, R., et al. (2001). Enhancement of the high-magnetic-field critical current density of superconducting MgB_2 by proton irradiation. *Nature, 411*, 561.

Corbette, J. P. (1990). Properties of boson-exchange superconductors. *Reviews of Modern Physics, 62*, 1027.

Das Sarma, S., Adam, S., Hwanf, E. H., & Rossi, E. (2011). Electronic transport in two-dimensional graphene. *Reviews of Modern Physics, 83*, 407–470.

Driessen, E. F. C., Coumou, P. C. J. J., Tromp, R. R., de Visser, P. J., & Klapwijk, T. M. (2012). Strongly disordered TiN and NbTiN s-wave superconductors probed by microwave electrodynamics. *Physical Review Letters, 109*, 107003.

Dwight, H. B. (1961). *Tables of integrals and other mathematical data*. MacMillan Co.

Eliashberg, G. M. (1960). Interactions between electrons and lattice vibrations in a superconductor. *Soviet Physics, JETP, 11*, 696 (in Russian, *Journal of Experimental and Theoretical Physics (USSR), 38*, 966, Mar. 1960).

Eliashberg, G. M. (1961). Temperature Green's function for electrons in a superconductor. *Soviet Physics, JETP, 12*, 1000 (in Russian, *Journal of Experimental and Theoretical Physics (USSR), 39*, 1437, Nov. 1960).

Fetter, A. L., & Walecka, J. D. (1971). *Quantum theory of many-particle systems*. McGraw-Hill.

Finkelstein, A. M. (1987). Superconducting transition temperature in amorphous films. *Pis'ma v Zhurnal Eksperimental'noi i Teoreticheskoi Fiziki, 45*, 37. *JETP Letters, 45*, 46.

Fisher, D. S., Fisher, M. P. A., & Huse, D. A. (1991). Thermal fluctuations, quenched disorder, phase transitions, and transport in type-II superconductors. *Physical Review B, 43*, 130.

Galitski, V. M., & Larkin, A. I. (2001). Disorder and quantum fluctuations in superconducting films in strong magnetic fields. *Physical Review Letters, 87*, 087001.

Ginzburg, V. L., & Kirzhnits, D. A. (Eds.). (1982). *High-temperature superconductivity*. New York: Consultants Bureau (A. K. Agyer, Trans., J. L. Birman, Ed.; original Russian text, Nauka, Moscow, 1977).

Karnaukhov, I. M., & Shepelev, A. G. (2008). Type II superconductors are 70 years old. *Europhysics News, 39*, 35.

Kemper, A., Doluweera, D. G. S. P., Maier, T. A., Jarrell, M., Hirschfeld, P. J., & Cheng, H.-P. (2009). Insensitivity of superconductivity to disorder in the cuprates. Retrieved from arXiv:0807.0195v2.

Kierfeld, J., & Vinokur, V. (2004). Lindemann criterion and vortex lattice phase transitions in type-II superconductors. *Physical Review B, 69*, 024501.

Kim, J. H., Dou, S. X., Oh, S., Jercinovic, M., Babic, E., Nakane, T., et al. (2008). Correlation between doping induced disorder and superconducting properties in carbohydrate doped MgB_2. Retrieved from arXiv:0810.1558.

Kittel, C. (1968). *Introduction to solid state physics*. New York: Wiley.

Kittel, C. (1987). *Quantum theory of solids*. New York: Wiley.

Kotliar, G., & Kapitulnik, A. (1986). Anderson localization and the theory of dirty superconductors. II. *Physical Review B, 33*, 3146.

Kozhevnikov, V. F., Van Bael, M. J., Vinckx, W., Temst, K., Van Haesendonck, C., & Indekeu, J. O. (2005). Surface enhancement of superconductivity in tin. *Physical Review B, 72*, 174510.

Krowne, C. M. (2011). Nanowire and nanocable intrinsic quantum capacitances and junction capacitances: Results for metal and semiconducting oxides. *Journal of Nanomaterials*, 160639.

Maleyev, S. V., & Toperverg, B. P. (1988). On high-T_c superconductivity and structural disorder. *Solid State Communications, 67*, 405.

Martinez, J. I., Abad, E., Calle-Vallejo, F., Krowne, C. M., & Alonso, J. A. (2013). Tailoring structural and electronic properties of RuO_2 nanotubes: Many-body approach and electronic transport. *Physical Chemistry Chemical Physics, 15*(35), 14715–14722.

Martinez, J. I., Calle-Vallejo, F., Krowne, C. M., & Alonso, J. A. (2012). First-principles structural & electronic characterization of ordered SiO_2 nanowires. *The Journal of Physical Chemistry C, 116*, 18973–18982.

Mondaini, F., Paiva, T., dos Santos, R. R., & Scalettar, R. T. (2008). Disordered two-dimensional superconductors: Role of temperature and interaction strength. *Physical Review B, 78*, 174519.

Nakhmedov, E., Alekperov, O., & Oppermann, R. (2012). Effects of randomness on the critical temperature in quasi-two-dimensional organic superconductors. *Physical Review B, 86*, 214513.

Osofsky, M. S., Hernández-Hangarter, S. C., Nath, A., Wheeler, V. D., Walton, S. G., Krowne, C. M., et al. (2016). Functionalized graphene as a model system for the two-dimensional metal–insulator transition. *Scientific Reports, 6*, 19939.

Osofsky, M. S., Krowne, C. M., Charipar, K. M., Bussmann, K., Chervin, C. N., Pala, I. R., et al. (2016). Disordered RuO_2 exhibits two dimensional, low-mobility transport and a metal–insulator transition. *Scientific Reports, 6*, 21836.

Pines, D. (1964). *Elementary excitations in solids: Lectures on phonons, electrons and plasmons*. New York: W. A. Benjamin, Inc. (First publ. 1963; corrections in 2nd printing, 1964).

Pines, D. (1979). *The many-body problem* (5th printing). *Frontiers in physics, a lecture note & reprint volume*. Reading, MA: Benjamin/Cummings Publ. Co.

Reif, F. (1965). *Fundamentals of statistical and thermal physics*. McGraw-Hill.

Schrieffer, J. R. (1983). *Theory of superconductivity*. *Frontiers in physics*. Reading, MA: Perseus Books (Orig. publ. 1964; 1983 ed. has Nobel lectures).

Siemons, W., Steiner, M. A., Koster, G., Blank, D. H., Beasley, M. R., & Kapitulnik, A. (2008). Preparation and properties of amorphous MgB_2/MgO superstructures: Model disordered superconductor. *Physical Review B*, 77, 174506.

Su, X., Zuo, F., Schlueter, J. A., Kelly, M. E., & Williams, J. M. (1998). Structural disorder and its effect on the superconducting transition temperature in the organic superconductor κ-(BEDT-TTF)$_2$Cu[N(CN)$_2$]Br. *Physical Review B*, *57*, 14056.

Swanson, M., Loh, Y. L., Randeria, M., & Trivedi, N. (2014). Dynamical conductivity across the disorder-tuned superconductor–insulator transition. *Physical Review X*, *4*, 021007.

Tanaka, K., & Marsiglio, F. (2000a). Possible electronic shell structure of nanoscale superconductors. *Physics Letters A*, *265*, 133.

Tanaka, K., & Marsiglio, F. (2000b). Anderson prescription for surfaces and impurities. *Physical Review B*, *62*, 5345.

Tinkham, M. (1980). *Introduction to superconductivity*. Malabar, FL: Krieger Publ. Co. (Orig. ed., McGraw-Hill, 1975).

Wang, Y. L., Wu, X. L., Chen, C.-C., & Lieber, C. M. (1990). Enhancement of the critical current density in single-crystal $Bi_2Sr_2CaCu_2O_8$ superconductors by chemically induced disorder. *Proceedings of the National Academy of Sciences of the United States of America*, *87*, 7058.

CHAPTER THREE

Mirror Electron Microscopy

A.B. Bok[1], J.B. le Poole✠, J. Roos, H. de Lang
Philips P.I.T.E.O., Eindhoven, Netherlands
TNO-TH, Delft, Netherlands
[1]Corresponding author: e-mail address: fred@familiebok.eu

Contents

✠ Deceased. Reprinted from A.B. Bok, J.B. Le Poole, J. Roos, and H. de Lang. Advances in Optical and Electron Microscopy 4 (1971) 161–261. With appendices by H. Bethge, J. Heydenreich, and M.E. Barnett.

Advances in Imaging and Electron Physics, Volume 203
ISSN 1076-5670
http://dx.doi.org/10.1016/bs.aiep.2017.07.005

1. INTRODUCTION

From the birth of electron optics (about 1920) till the late 1950s most efforts in the field of electron optics were concentrated on the theory, the design and development of the nowadays widely available transmission electron microscope. A guaranteed point resolution of less than 0.5 nm is considered normal for high quality instruments.

Since the transmission electron microscope provides information about the internal structure of an electron transparent specimen, this technique does not allow for the direct investigation of surfaces of solids. Two useful alternatives are either putting both the illuminating and imaging system at a glancing angle with the surface to be examined (reflection electron microscopy) or the application of the replica technique. The indirect observation of a surface by means of a replica permits a resolving power up to 5 nm.

The increasing interest in direct observation of surfaces of solids or investigation of surface phenomena has resulted during the past 15 years in the development of the following types of electron microscopes:

1. scanning electron microscope;
2. emission electron microscope;
3. reflection electron microscope;
4. mirror electron microscope.

Before going into more detail concerning the mirror electron microscope a brief description of the other types of microscopes is presented.

1.1 Scanning Electron Microscope

In a scanning electron microscope (Knoll, 1935) a primary electron beam, emitted from a heated tungsten filament, is focused into a fine electron probe on the specimen and made to scan on a raster—similar to television techniques—on the surface by a deflection system. Electrons liberated from the specimen by the focused primary beam are detected by a photomultiplier tube with a scintillator mounted on top. The photomultiplier output signal is used to modulate the brightness of the electron beam in a cathode-ray tube, which is scanned in synchronism with the electron probe. The resolution—being of the order of 20 nm in favorable operating conditions—depends upon the diameter of the electron probe, the accelerating voltage, the detector system and the type of specimen.

1.2 Emission Electron Microscope

In an emission electron microscope the specimen acts as a self-illuminating object. Electrons are liberated from the specimen by either heating of the specimen (thermionic emission), electron and ion bombardment of the specimen (secondary emission) or quantum irradiation of the specimen (photo-emission). The image is usually formed by a combination of two or three electron lenses. The obtainable resolution—mainly determined by the energy spread of the emitted electrons and the strength of the electrostatic field at the specimen surface—amounts to about 20 nm. In the case of thermionic and photo emission the image contrast is mainly dominated by the local work function of the specimen surface. Since the successful application of photo-emission, by means of ultra-violet radiation, this type of microscope has become of great importance.

1.3 Reflection Electron Microscope

Reflection electron microscopy is rarely used nowadays. The first experiments (Ruska, 1933) did not show very promising results until von Borries and Jansen (1941) suggested that the large energy spread of the scattered electrons could be reduced by having the illuminating and imaging system at a glancing angle with the specimen. The remaining energy spread still requires a small aperture in order to minimize the dominating chromatic aberrations. Since the reflected electrons are scattered over a wide angle, a small angular aperture of the accepted beam has to be selected. This gives an image barely bright enough to be focused at the necessary magnification.

1.4 Mirror Electron Microscope

Contrary to the techniques mentioned above the specimen is neither struck by electrons nor emits electrons. An accelerated electron beam enters the retarding field of an electrostatic mirror. Application to the mirror electrode of a potential, which is slightly more negative than the accelerating voltage, causes the electrons to be reflected from an equipotential plane closely in front of the mirror electrode, which is in this technique the specimen surface. The electron trajectories near the point of reversal, in front of the specimen, are highly sensitive to deviations from flatness of the reflecting equipotential plane. These deviations are either caused by electrostatic or topographic perturbations at the physical specimen surface.

The possibility of converting on a microscopic scale electrostatic and, to a certain amount, magnetic potential distributions into a directly observable image has given access to new information in phenomena such as diffusion of metals, contact potentials, surface conductivity and magnetic properties.

The first experimental results of Hottenroth (1937) and the calculations of Recknagel (1936) and Henneberg and Recknagel (1935) clearly revealed the feasibility of mirror electron microscopy. Hottenroth showed that the manner of formation of non-focused images of the mirror electrode closely resembles that of the light optical "Schlieren" method. Following his experiments the research in this field was mainly directed toward the application of this technique for visual observation of surface phenomena. Numerous articles, especially by Mayer (1957a, 1957b, 1959a, 1959b), Spivak, Prilegaeva, and Azovcev (1955), Spivak, Igras, Pryamkova, and Zheludev (1959a, 1959b), Igras (1961), Spivak (1959), and Spivak, Saparin, and Pereversev (1962), are published about different kinds of applications with this type of microscope.

Little attention has been paid to optimizing the imaging technique of the mirror electrode. It was Le Poole (1964c) who pointed out that the attainable resolving power for this type of microscope could be improved considerably by forming a focused image of the mirror electrode onto the fluorescent screen. Also it became evident that mirror electron microscopes with rotationally symmetric lenses require a separation of the illuminating and reflected beam in order to obtain a focused image of the specimen with sufficient field of view. In instruments with rotationally symmetric lenses and without beam separation (Bethge, Hellgardt, & Heydenreich, 1960; Forst & Wende, 1964; Barnett & Nixon, 1967a, 1967b) the specimen is illuminated through a central hole in the final screen. The field of view, which disappears entirely when the mirror electrode is exactly conjugated to the final screen, can only be increased by defocusing. The formation of contrast in a microscope with focused images is achievable in a way similar to the transmission electron microscope by having an aperture in the objective lens. To avoid a new limitation of the field of view and to have at the same time normal incidence of the illuminating beam this aperture must be in the back focal plane of the objective lens. Simultaneous focusing of the specimen and a sufficiently large field of view can be obtained by the separation of the illuminating and reflected beam with a magnetic prism or the use of magnetic quadrupoles.

Concerning this latter method a scanning mirror electron microscope with magnetic quadrupoles has been designed and constructed (Bok, Kramer, & Le Poole, 1964). A brief description of this instrument is given in Appendix A.

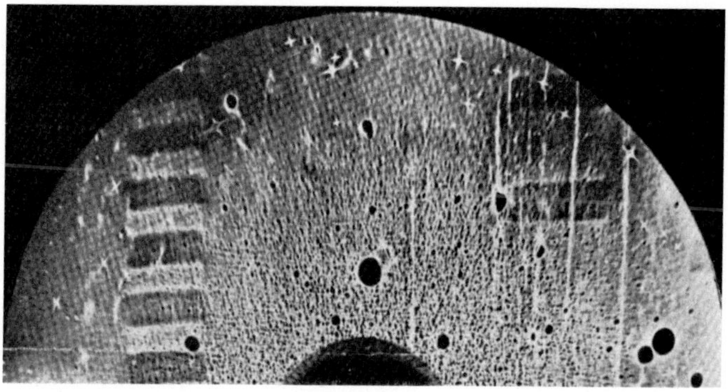

Figure 1 Mirror projection image of a magnetic recording pattern (made by L. Mayer).

Although beam separation with a magnetic prism has been applied earlier by several experimenters (Orthuber, 1948; Bartz, Weissenberg, & Wiskott, 1956; Hopp, 1960; Schwartze, 1967), it was never used (except for the instrument of Schwartze, 1967) in combination with a contrast aperture in the back focal plane of the objective lens.

The absence of this aperture means that for in-focus images the contrast disappears. Here defocusing provides the necessary image contrast.

From geometrical considerations it follows that for an imaging mirror microscope special attention has to be paid to the condenser system. Seen from the objective aperture the electron source must appear as large as the required field of view.

This leads to the conclusion that all mirror electron images, obtained so far, except for some pictures made by Schwartze (1967) are point projection, out of focus, images.

When a point projection image of the mirror electrode is to be made, an electron probe is formed in front of the specimen. The electrons then reflect from a paraboloid of revolution which is the envelope of all parabolic electron trajectories in the retarding field. Contrary to this, in a microscope with focused images and an objective aperture in the back focal plane, all electrons reflect from a flat equipotential plane normal to the z-axis.

The effect of the reflecting paraboloid in defocused instruments is clearly observable from most of the photographic results published. Where electrons strike the specimen surface local negatively charged spots occur. Negative spots give rise to black "bubbles" in the final image. For positively charged spots, caused for instance by a positive ion bombardment on areas where the electrons do not reach the specimen surface, white "stars"

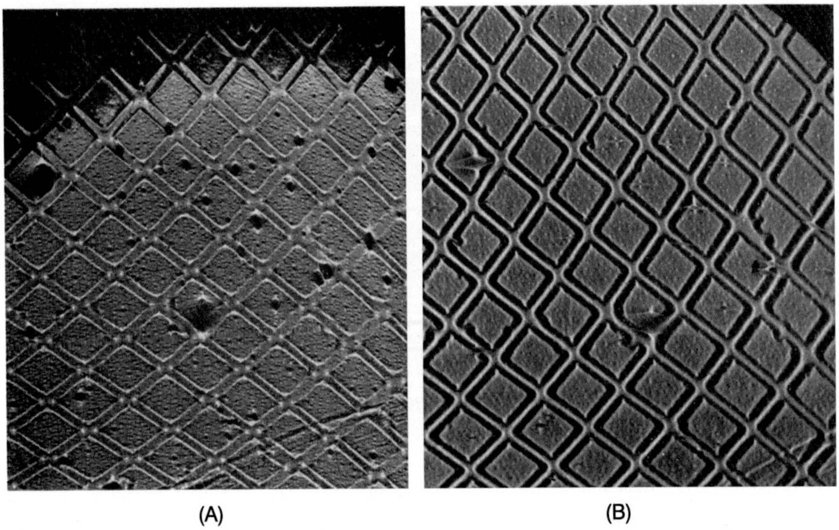

Figure 2 Gold squares, about 20 mm thick, vacuum deposited through a 750 mesh grid on a layer of gold (magnification ×330). (A) Slightly under focus. (B) Slightly over focus.

emerge on the final screen (Lenz & Krimmel, 1963) (see Fig. 1). When, in a mirror projection microscope, the electrons with the highest energy in the Maxwellian distribution and incident close to the axis are allowed to strike the specimen, the central region on the final screen shows mainly black bubbles. The more off axis electrons reverse their direction before reaching the specimen surface and give mainly white stars for the outer regions (see Fig. 1). On the other hand, the occurrence of some stars in the central region and bubbles in the outer regions is comprehensible owing to the fact that, apart from local charges, the topography of the surface also gives rise to similar effects. In-focus images show hardly any black bubbles and white stars. Near the focusing condition, where the contrast reverses, the bubbles change into stars and vice versa (see Figs. 2A and B).

The main subject of this article is the theory, design and construction of a mirror electron microscope, suitable for focused images. It consists of rotationally symmetric lenses, a magnetic prism, a wide aperture condenser system and a contrast aperture in the back focal plane of the objective lens. In Section 2 a treatment is given of the formation of contrast in a mirror electron microscope, whereas Section 3 deals with the design and construction of the instrument. Section 4 provides a series of photographic results and a brief discussion about possible applications.

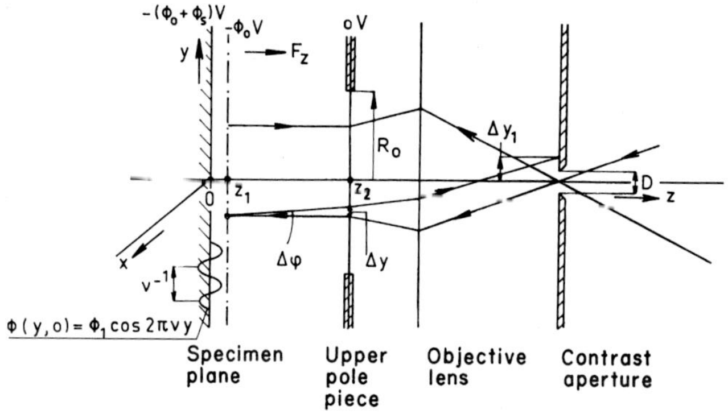

Figure 3 Retarding field with perturbed specimen and characteristic quantities.

2. CONTRAST FORMATION IN A MIRROR ELECTRON MICROSCOPE WITH FOCUSED IMAGES[1]

2.1 Principle

When a mono-energetic and axially parallel beam of electrons enters a homogeneous electrostatic retarding field, reflection occurs against a flat equipotential plane normal to the z-axis (see Fig. 3) and all electrons return along the same trajectories.

Owing to local deviations from flatness of the reflecting equipotential plane, the electrons which approach these perturbations receive a tangential impulse. These electrons describe a different trajectory after reversal and intersect the aperture plane at a height Δy_1, which depends on the perturbation present. When Δy_1 exceeds $D/2$, the radius of the contrast forming aperture, the electrons impinge on the aperture and are removed from the reflected electron beam. The formation of contrast in a mirror electron microscope resembles the technique (an aperture in the back focal plane of the objective) applied in transmission electron microscopes or the optical "Schlieren" technique. The separation of the tangentially modulated electron pencils from the unperturbed pencils allows for visual observation of perturbations in the reflecting equipotential plane, in terms of current density modulations in the final image. The origin of these perturbations can be twofold, topography of an equipotential specimen surface or electrostatic disturbances on a flat specimen surface. In practice, mostly a combination of both is encountered.

[1] See Section 2.8 for a concise list of symbols used in Section 2.

The current density modulations in the final image do not provide direct information about the type of perturbation present at the specimen. Perturbations at the specimen of magnetic origin, in comparison with electrostatic and topographic perturbations, hardly affect an axially parallel beam of electrons. Magnetic contrast will be discussed separately in Section 2.7.

2.2 Calculations

The purpose of the calculations given in Sections 2.3, 2.4, and 2.5—restricted to electrostatic and topographic contrast—is to find the dependence of the lateral shift Δy_1 or the related angular deflection $\Delta\varphi_1$ of a reflected electron pencil on the specimen perturbation.

All three sections assume a mono-energetic ($e\phi_o$) and axially parallel beam of electrons hi the homogeneous retarding field.

In Fig. 3 the specimen coincides with the *xoy*-plane. The calculations are only performed in the *yoz*-plane. The electrostatic retarding potential $(\phi_o + \phi_s)$, a superposition of the accelerating voltage ϕ_o and an additional voltage ϕ_s, is considered homogeneous and unaffected by the bore in the upper pole piece. This is valid provided that z_2 is at least three times the bore radius R_o (Glaser, 1952). This assumption permits the separation of the divergent lens action of the upper pole piece from the contrast formation mechanism near the specimen. Apart from lens defects the divergent lens action does not affect the contrast formation. It only requires a slightly higher excitation of the objective lens to maintain the parallel incidence into the mirror field. The behavior of electrons in the retarding field can be described either classically by the equations of motion (1a) and (1b) or wave mechanically by the time independent Schrödinger equation (2).

Classically

y-direction

$$-e\frac{\partial\Phi(y,z)}{\partial y} = -e\frac{\partial\phi(y,z)}{\partial y} = m\frac{d^2y}{dt^2} \tag{1a}$$

z-direction

$$-e\frac{\partial\Phi(y,z)}{\partial z} = -eF_z - e\frac{\partial\phi(y,z)}{\partial z} = m\frac{d^2z}{dt^2} \tag{1b}$$

Wave mechanically

$$-\frac{\hbar^2}{2m}\left(\frac{\partial^2 u}{\partial y^2} + \frac{\partial^2 u}{\partial z^2}\right) + (V - E)u = 0 \tag{2}$$

where

$$\Phi(y, z) = \phi_o + \phi_s + \phi(y, z) \tag{3}$$

$e =$ elementary charge

$m =$ electron rest mass

$\phi(y, z)$ is the perturbation potential

F_z is the strength of the retarding field.

For 30 kV across a gap of 3.5×10^{-3} m, $F_z = 8.57 \times 10^6$ V/m.

$\hbar = \frac{h}{2\pi}$, $h =$ Planck's constant

$u = u(y, z)$ the wave function

$V = V(z)$ the potential energy in the retarding field

E is the kinetic energy of the incident beam.

For all calculations following it is assumed that $\phi_s \ll \phi_o$.

Two models A and B, Section 2.3 and 2.4, are based on the equations of motion (1a) and (1b), whereas model C (Section 2.5) uses the Schrödinger equation (2).

It would be obvious to describe the formation of contrast in a way comparable with the modulation transfer functions in the light optics. The non-linear character of Eqs. (1a) and (1b), however, does not allow for such a description because no linearity exists in the case of sufficient contrast between the perturbation amplitude at the specimen and the tangentially modulated electron pencils. Since it is wished to provide an analytical description of the contrast mechanism, preferably in a way resembling the modulation transfer functions, Eqs. (1a) and (1b) are linearized by the assumption $\frac{\partial \phi(y,z)}{\partial z} \ll F_z$ (model A). This causes a sinusoidal specimen to produce a sinusoidal modulation.

Contrary to the approximated model A, model B provides information about the solution of the exact equations (1a) and (1b). The calculations for this model were performed both on a digital computer (Telefunken TR-4) and an analog computer (Applied Dynamics 40). The TR-4 calculations provide the Δy value and the corresponding coordinates of the point of reversal for different heights of incidence. The omitted index 1 for Δy and

$\Delta\varphi$ indicates that these values are measured in the plane $z = z_2$, the upper pole piece of the objective lens.

A comparison of the results obtained with models A and B is shown to lead for certain values of the local slope of the reflecting plane to a matching of both models. This means that the non-linear model B shows a linear behavior.

As $\phi(y, z)$ fulfills the Laplace equation $\Delta\phi(y, z) = 0$ a sinusoidal perturbation potential

$$\phi(y, 0) = \phi_1 \cos 2\pi n\nu y \text{ causes a potential}$$
$$\phi(y, z) = \phi_1 \cos 2\pi n\nu y \exp(-2\pi n\nu z),$$

where ν is the spatial specimen frequency and ϕ_1 the perturbation amplitude.

Since the assumption $\frac{\partial\phi(y,z)}{\partial z} \ll F_z$ involves simultaneously moderate values $\frac{\partial\phi(y,z)}{\partial y}$, it can be expected prior to the calculations following that models A and B match only for specimens slightly perturbed (small values of $\phi_1\nu$). Specimens with more contrast are not accessible to a simple analytical description. In that case numerical calculations should provide information.

Figs. 4A and B, both made on an analog computer, demonstrate the effect of the assumption $\frac{\partial\phi(y,z)}{\partial z} \ll F_z$ the electron trajectories near a sinusoidal perturbed specimen. In model A all electrons reflect from a flat equipotential plane whereas in model B $\frac{\partial\phi(y,z)}{\partial z}$ leads to a variable depth of penetration into the retarding field.

Apart from Eqs. (1a) and (1b) a second non-linear effect in the formation of contrast is introduced by the filtering of tangentially modulated electron pencils from the reflected beam with a circular contrast aperture.

Since this effect is independent of the sign of the lateral deflection, the modulated current density distribution in the final image shows the double frequency of a sinusoidal perturbation. This double rectifying effect can be avoided by using a knife edge as aperture or, in the case of two dimensional specimens, two perpendicular edges.

In order to avoid, at this stage, the choice between the non-linearly filtering circular aperture and the linearly operating edge aperture, the lateral displacement Δy and the angular deflection $\Delta\varphi$ are both plotted against the spatial perturbation frequency and amplitude.

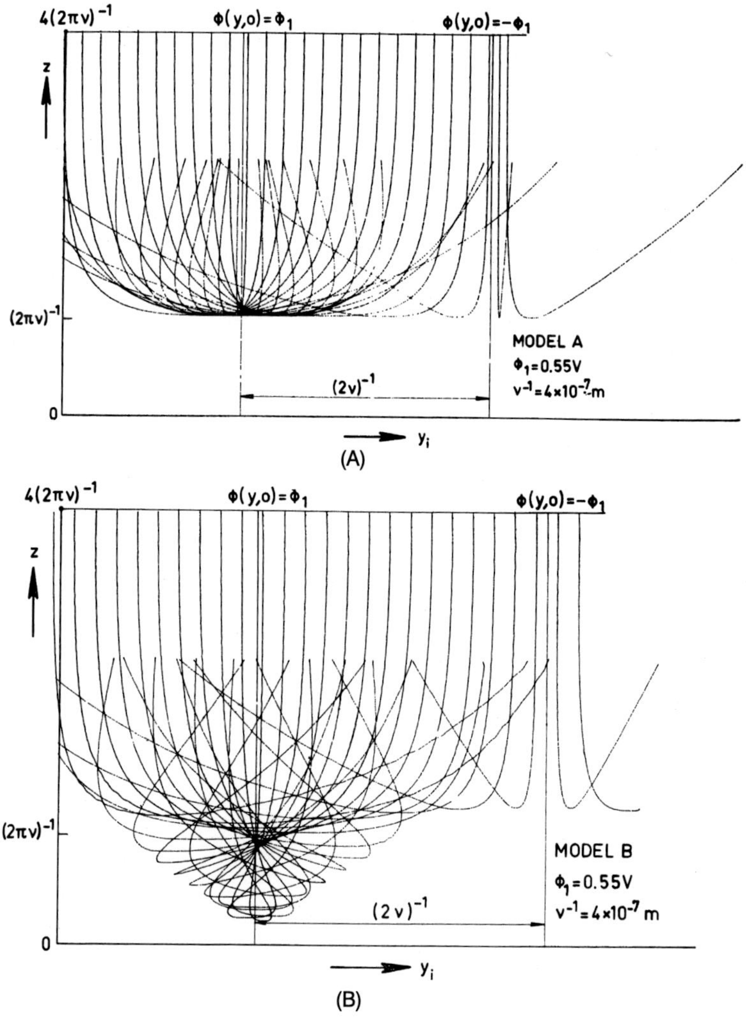

Figure 4 (A) Electron trajectories according to model A. (B) Electron trajectories according to model B.

2.3 Model A

2.3.1 Electrostatic Contrast

The simplified equations of motion are

$$-e\frac{\partial\Phi(y,z)}{\partial y} = m\frac{d^2y}{dt^2} \tag{4a}$$

$$-eF_z = m\frac{d^2z}{dt^2} \tag{4b}$$

where

$$\Phi(y,z) = \phi_o + \phi_s + \phi(y,z); \quad \phi_s = \phi_{SC} + \phi_{SV} \tag{4c}$$

Superimposed on the retarding potential ϕ_o is a "specimen" voltage ϕ_s. This additional negative voltage prevents electrons from striking the specimen near positively charged perturbations. ϕ_s, which is the sum of ϕ_{SC}, the contact potential between the specimen material and the tungsten filament in the electron gun, and ϕ_{SV}, a variable voltage, determines the distance z_1 of the reflecting equipotential plane in front of the specimen surface.

If ϕ_{SC} is corrected for then

$$z_1 = \frac{\phi_{SV}}{F_z} \ll z_2$$

In this linearized model it is useful to represent an electrostatic (or topographic) perturbation $\phi(y, 0)$ along the y-axis as a Fourier series (or integral).

$$\phi(y,0) = \sum_{n=1}^{\infty} \phi_n \cos 2\pi n\nu y \tag{5}$$

The omitted term with $n = 0$, an additional voltage on top of the specimen potential, is defined as ϕ_{SA}.

In these calculations no incident electrons are allowed to reach the physical specimen surface, because the electron scattering effects which would occur, destroy the validity of the results obtained. Experiments revealed, that for a specimen biased slightly positive (some tenth of volts) with respect to the accelerating voltage ϕ_o, the image contrast deteriorated considerably due to the electron scattering phenomena at the specimen surface.

The lateral impulse Δmv_y given an electron traveling toward and from the specimen amounts to

$$\Delta mv_y = -2\int_{z_1}^{z_2} e\frac{\partial\phi(y,z)}{\partial y}dt \approx -2\int_{o}^{\infty} e\frac{\partial\phi(y,z)}{\partial y}dt \tag{6}$$

Inserting Eqs. (4c) and (5) into (6) and neglecting the lateral displacement during reversal, which is permissible for small values of Δmv_y, it follows that:

$$\Delta m v_y = e \sum_{n=1}^{\infty} [\phi_n 2\pi n\nu \exp(-2\pi n\nu z_1) \sin 2\pi n\nu y]$$
$$\times 2 \int_o^{\infty} \exp[-2\pi n\nu (z - z_1)] dt$$

$$\Delta m v_y = \left(\frac{4\pi^2 me}{F_z}\right)^{\frac{1}{2}} \sum_{n=1}^{\infty} [\phi_n (n\nu)^{\frac{1}{2}} \sin 2\pi n\nu y \exp(-2\pi n\nu z_1)] \quad (7)$$

ϕ_{SV} has been introduced by writing $(z - z_1)$ instead of z.

If y_i represents the height of incidence above the z-axis and y_r the corresponding height for the reflected electrons, both measured in the plane $z = z_2$, then

$$y_r - y_i = \Delta y|_{z=z_2} = v_y t$$

$$\Delta y|_{z=z_2} = \frac{2\pi}{F_z} (2_{z_2} \nu)^{\frac{1}{2}} \sum_{n=1}^{\infty} [\phi_n n^{\frac{1}{2}} \sin 2\pi n\nu y \exp(-2\pi n\nu z_1)] \quad (8)$$

After interaction with the specimen the reflected electrons follow parabolic trajectories in the retarding field, owing to the lateral momentum received.

The corresponding angular deflection $\Delta\varphi$ is

$$\Delta\varphi = \frac{\Delta y}{2z_2} \quad (9)$$

For $n = 1$, Δy and $\Delta\varphi$ are determined as a function of ϕ_1 and ν, the spatial frequency of a sinusoidal perturbation at the specimen surface,

$$\Delta y|_{z=z_2} = \Delta y_\nu = \frac{2\pi}{F_z} (2z_2\nu)^{\frac{1}{2}} \phi_1 \sin 2\pi \nu y \exp\left(-2\pi - \frac{\phi_1}{\phi_o} z_2 \nu\right)$$
$$\times \exp\left(-2\pi \frac{\phi_{SA}}{\phi_o} z_2 \nu\right) \quad (10)$$

The index ν in Δy_ν and $\Delta\varphi_\nu$ refers to electrostatic contrast in $z = z_2$. $z_1 = \frac{\phi_1 + \phi_{SA}}{\phi_o} z_2$ (with ϕ_{SC} corrected for) has a minimum value $\frac{\phi_1}{F_z}$ which prevents electrons in the case of electrostatic contrast from striking the specimen near positively charged perturbations.

The difference $\phi_{SV} - \phi_1 = \phi_{SA}$ corresponds to an additional voltage for adjusting the reflecting equipotential plane away from the perturbed specimen surface.

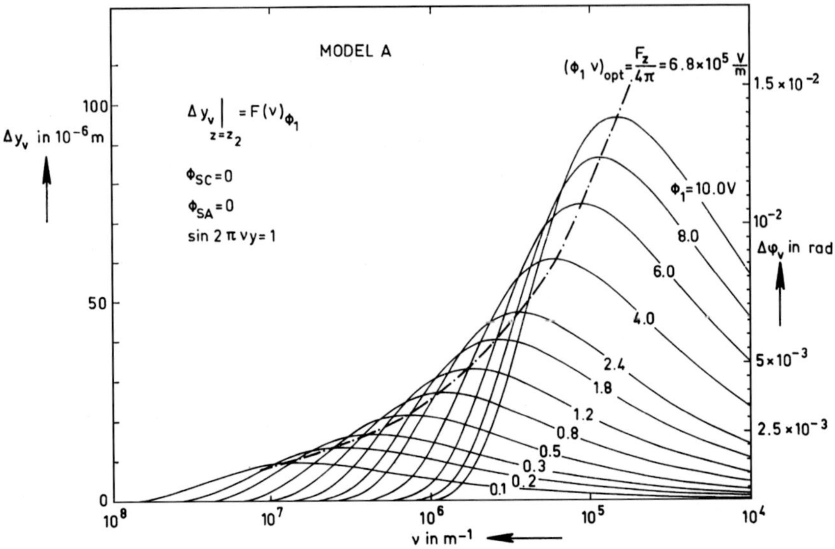

Figure 5 Modulation functions Δy_ν and $\Delta\varphi_\nu$ plotted against ν (model A).

In Fig. 5 the maximum values of Δy_ν and $\Delta\varphi_\nu$, following from Eq. (10), are plotted against ν with ϕ_1 as curve parameter. For all curves presented it is assumed that $\sin 2\nu y = 1$, $\phi_{SC} = \phi_{SA} = 0$. Since the local "slope" $\phi_1\nu$ of the perturbations in the reflecting equipotential plane plays the main role in this contrast mechanism, Δy_ν is also plotted against $\phi_1\nu$ (see Fig. 6).

The maxima of the plotted Δy_ν and $\Delta\varphi_\nu$ values for each curve $F(\phi_1\nu)_{\phi_1}$ coincide with the straight line $\frac{(\phi_1\nu)_{opt}}{F_z} = 8 \times 10^{-2}$ or in this instrument $(\phi_1\nu)_{opt} = \frac{F_z}{4\pi} = 6.8 \times 10^5$ V/m.

2.3.2 Topographic Contrast

Provided that $\frac{\partial d(y,z)}{\partial y} \ll 1$ the topographic "displacement" Δy_t is found by substituting in Eq. (8)

$$\phi_n = d_n F_z \quad \text{and} \quad z_1 = \frac{\phi_{SA}}{F_z} \tag{11}$$

These relations are only applicable for small perturbations with moderate curvatures ($\frac{\partial^2 d(y,z)}{\partial y^2} \ll 1$) because then the z component of the field strength near the specimen equals F_z.

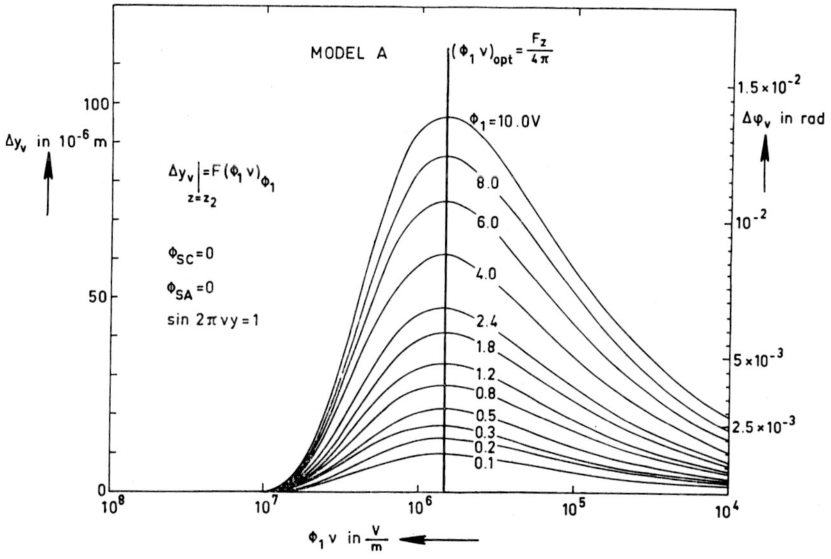

Figure 6 Modulation functions Δy_V and $\Delta\varphi_V$ plotted against $\phi_1 \nu$ (model A).

d_n represents the amplitude of the component harmonics in the specimen topography

$$d(y, 0) = \sum_{n=1}^{\infty} d_n \cos 2\pi n\nu y \tag{12}$$

Contrary to electrostatic contrast all topographic perturbations coincide with one equipotential plane. It follows that the minimum required bias for electrostatic contrast (for $n = 1$ is $z_1 = \frac{\phi_1}{F_z}$) can be omitted for topographic contrast.

The electrostatic contrast calculations of Barnett and Nixon (1967a, 1967b) have apparently overlooked this effect. Contrary to the title of their publication, it means that they have in fact calculated topographic contrast.

If $\phi_{SA} = 0$ and $n = 1$ all electrons reach exactly the "topographic specimen" and

$$\Delta y_t|_{z=z_2} = 2\pi (2z_2\nu)^{\frac{1}{2}} d_1 \sin 2\pi y \tag{13}$$

A similar plot to Fig. 5 presents the maximum values of Δy_t as a function of ν and d_1 with $\sin 2\pi\nu y = 1$, $\phi_{SA} = 0$ and $\phi_{SC} = 0$ (see Fig. 7).

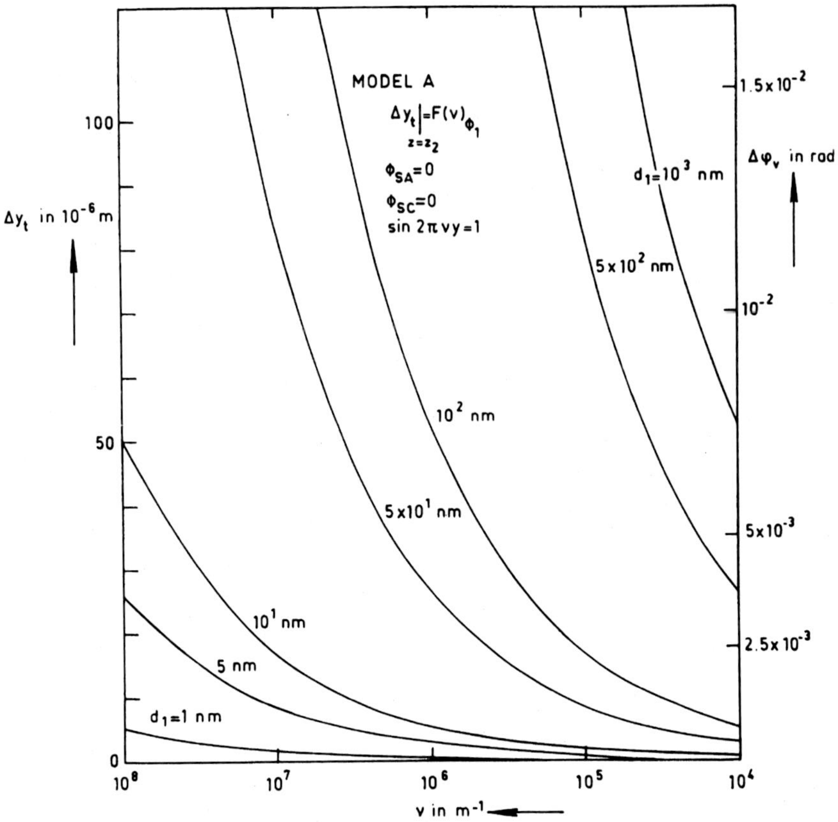

Figure 7 Modulation functions Δy_t and $\Delta\varphi_t$ plotted against ν (model A).

2.3.3 Conclusions and Remarks for Model A

Conclusions

1. An electrostatic or topographic cosine perturbation at the specimen surface gives a sine modulation on the angle of the reversing electron pencils. For electrostatic contrast the specimen should be at least biased with an additional negative voltage, equal to the positive amplitude of the perturbation signal. This prevents electrons from reaching the specimen surface.

2. It follows directly from the linear character of Eqs. (4a) and (4b) that the modulation effect of an arbitrary periodic perturbation, either electrostatic or topographic, can be calculated by summing the separate modulation effects of the component harmonics.

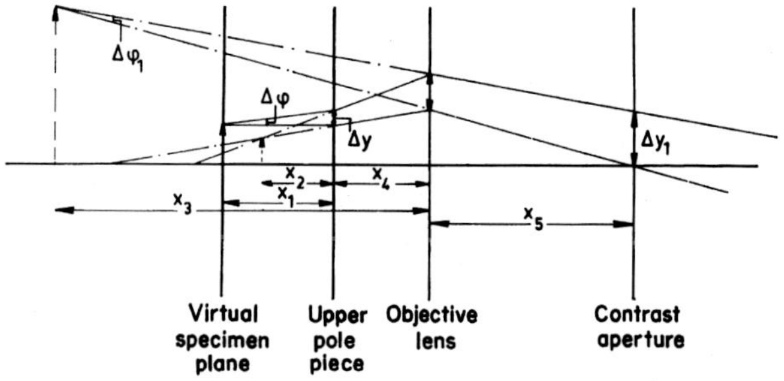

Figure 8 Relation between Δy on the upper pole piece and Δy_1 in the contrast aperture piano.

3. For topographic contrast with $\frac{\partial d(y,0)}{\partial y} \ll 1$ and $\frac{\partial^2 d(y,0)}{\partial y^2} \ll 1$ a linear relation exists between the perturbation amplitude and the modulation effect (Δy_t or $\Delta\varphi_t$) on the angle of the reversing electron pencils. A similar linear relation is valid for electrostatic contrast provided that $2\pi\phi_1\nu \ll F_z$.

4. A discrimination between topographic and electrostatic contrast is possible in principle, because for topography the modulation function is independent of the position at the specimen. For electrostatic contrast it varies from place to place owing to the additional damping factor.

5. The fact that $\phi_1\nu$ possesses a constant value corresponding to the maxima of the plotted Δy_ν or $\Delta\varphi_\nu$ values (Figs. 5 and 6) means that once ϕ_1 is given, the optimum ν value for maximum electrostatic contrast follows immediately.

The conclusions stated above involve the assumptions:

1. the illuminating beam in the retarding field is parallel to the z-axis;
2. $\frac{\partial\phi(y,z)}{\partial z} \ll F_z$;
3. $\frac{\partial d(y,0)}{\partial y} \ll 1$ and $\frac{\partial^2 d(y,0)}{\partial y^2} \ll 1$ for topographic contrast;
4. electrons do not reach the specimen surface;
5. the illuminating beam is mono-energetic ($e\phi_o$).

Remarks

1. The current density distribution in the final image can be determined by projecting the Δy value found onto the contrast aperture plane. For the combination of objective and electrostatic lens (see Fig. 8) the lateral shift Δy_1 in the contrast aperture plane amounts to

$$\Delta y_1 = \left(1 + \frac{x_4}{x_2}\right)\left(1 + \frac{x_5}{x_3}\right)\Delta y = G_1 \Delta y$$

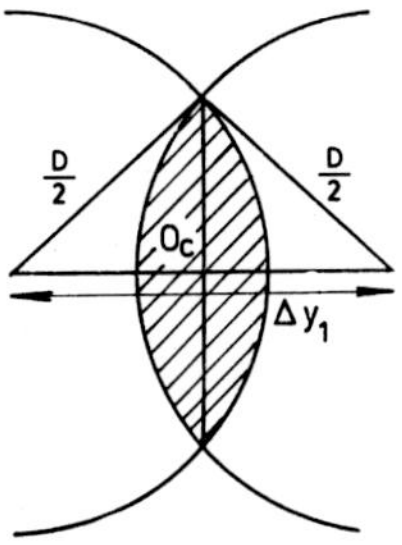

Figure 9 Common area of contrast aperture and reflected electron pencil.

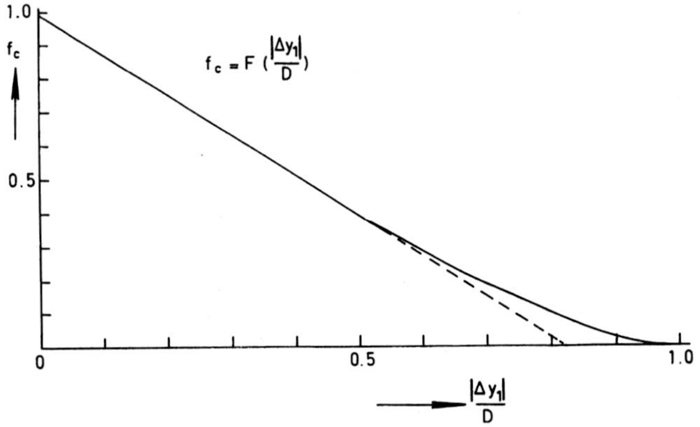

Figure 10 Filtering characteristic f_C for a "filled" circular contrast aperture.

This expression holds for both the virtual and the real imaging mode of the objective lens.

Every point of the specimen surface inside the field of view is illuminated by an electron pencil incident parallel to the z-axis. When a circular contrast aperture is used and the illuminating pencils fill this aperture, the common area 0_c of the contrast aperture and the cross section of each lateral deflected electron pencil directly indicates the current density in the final image. From Fig. 9 is found

$$0_c = \frac{1}{2}D^2\left\{\arccos\frac{|\Delta y_1|}{D} - \frac{|\Delta y_1|}{D}\left[1-\left(\frac{\Delta y_1}{D}\right)^2\right]^{\frac{1}{2}}\right\} \tag{14}$$

Plotting $f_c = \frac{0_C}{\frac{1}{4}\pi D^2}$ against $\frac{|\Delta y_1|}{D}$ shows an almost proportional relation (see Fig. 10).

Starting from a sinusoidal perturbed specimen

$$\phi(y,0)=\phi_1 \cos 2\pi\nu y \quad \text{for electrostatic contrast}$$

and

$$d(y,0)=d_1 \cos 2\pi\nu y \quad \text{for topographic contrast,}$$

the corresponding current density distribution $j(y)$, calculated back on the specimen surface, follows from

$$j(y)=j_o f_c \tag{15}$$

where j_o = current density of the illuminating beam without perturbations at the specimen and

$$f_c=\frac{2}{\pi}\left\{\arccos\frac{G_1|\Delta y_1|}{D}-\frac{G_1|\Delta y_1|}{D}\left[1-\left(\frac{G_1\Delta y_1}{D}\right)^2\right]^{\frac{1}{2}}\right\}\phi_{SA}=\phi_{SC}=0$$

and

$$\Delta y=\Delta y_v|_{z=z_2}=\frac{2\pi}{F_z}(2z_2\nu)^{\frac{1}{2}}\phi_1 \sin 2\pi\nu y \exp(-2\pi\nu v_1) \tag{8}$$

for electrostatic contrast or

$$\Delta y=\Delta y_t|_{z=z_2}=2\pi(2z_2\nu)^{\frac{1}{2}}d_1 \sin 2\pi\nu y$$

for topographic contrast.

Another filtering characteristic f_c would be

$$f_c=1 \quad \text{for } 0\leq G_1|\Delta y|\leq D/2$$

and

$$f_c=0 \quad \text{for } G_1|\Delta y|>D/2$$

It assumes a cross-over for the illuminating electron beam small compared with D.

Since D is restricted in practice to a lower limit of 50 μm, the crossover of the illuminating beam will neither fill the contrast aperture entirely nor has a size which makes it much smaller than D. The shape of the practical f_c curves for a circular aperture resembles a trapezium with the edges rounded off.

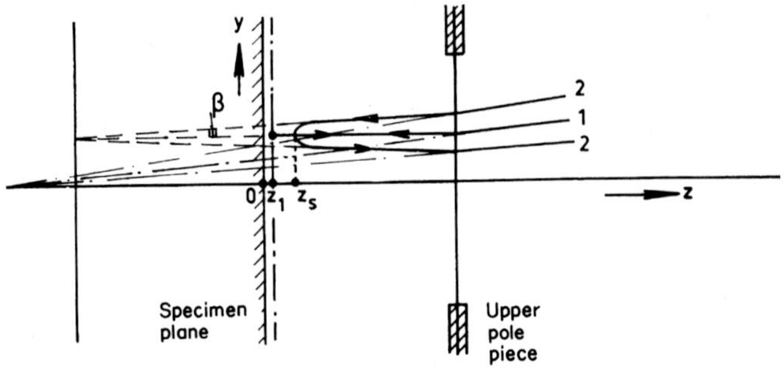

Figure 11 Spherical aberration of the retarding field.

For a knife edge aperture and a small cross-over

$$f_c = 1 \quad \text{for } G_1 \Delta y < 0$$
$$f_c = 0 \quad \text{for } G_1 \Delta y > 0$$

2. The assumption of specimen illumination with an electron beam perfectly parallel to the z-axis requires an infinitely small semi-angular aperture β of the composing pencils.

However, in order to fill the aperture, β needs to have a fairly large value. The value of β, depending on the circular contrast aperture D, follows from Fig. 8 where

$$\Delta\varphi_1 = \beta \quad \text{for } \Delta y_1 = D/2$$

$$\beta = \frac{D}{2x_1}\left[\left(1+\frac{x_5}{x_3}\right)\left(1+\frac{x_4}{x_2}\right)\right]^{-1} = \frac{D}{2x_1 G_1} \tag{16}$$

The effect of $\beta \neq 0$ is twofold:

(a) Owing to the spherical aberration of the homogeneous retarding field, electrons reflect against equipotential planes at $z = z_s(\beta)$ instead of $z = 0$ ($\phi_{SA} = \phi_{SC} = 0$).

The electron trajectories drawn in Fig. 11 correspond to one electron traveling parallel to the z-axis, having its point of reversal in $z = 0$, and a second electron entering the retarding field at an angle β with a point of reversal in $z = z_s$. The difference in depth of penetration amounts to

$$z_s = \beta^2 z_2 \quad \text{or in terms of voltage} \quad \phi_{SA} = \beta^2 \phi_o \tag{17}$$

Here ϕ_{SA} represents the voltage equivalent of z_s.

(b) Not derivable from this simplified contrast model are the high sensitivities of Δy_t and Δy_v for electrons approaching the perturbed reflecting equipotential plane at an angle with the z-axis. This effect, being more predominant than the damping effect $\phi_{SA} = \beta^2\phi_o$, is examined both with a digital and an analog computer on the basis of the exact equations (1a) and (1b).

See for further results Fig. 16B and C of Section 2.4 and Section 2.5.

3. As can be learned from Eq. (8)

$$\Delta y_v|_{z=z_2} = \frac{2\pi}{F_z}(2z_2\nu)^{\frac{1}{2}} \sum_{n=1}^{\infty}\left[\phi_n \sin 2\pi n\nu y \exp(-2\pi n\nu z_1)n^{\frac{1}{2}}\right]$$

a strongly increasing damping $n^{\frac{1}{2}} \exp(-2\pi n\nu z_1)$ occurs in the case of electrostatic contrast for higher order harmonics. Eq. (8) corresponds to a low pass filter. It causes structures with higher order harmonics to be hardly distinguishable from a sinusoidal perturbation.

4. The assumption of a mono-energetic beam of electrons is not feasible in practice owing to the Maxwellian energy distribution for electrons emitted from a heated filament. As a result of the Boersch effect, the retarded electron beam in front of the specimen shows even a larger spread in energy than following from the Maxwellian distribution. The corresponding variations in depth of penetration lead to a decrease in contrast because of the exponential z damping in the contrast modulation function.

The contribution of electrons with an energy between E_1 and $E_1 + \Delta E_1$ (after passing the retarding potential ϕ_o) to the formation of image contrast can be determined by multiplying the exponential z damping factor (curve 2 in Fig. 12).

$$\exp(-2\pi\nu z) = \exp\left[-\frac{2\pi\nu}{eF_z}(e\phi_{SV} - E_1)\right] \tag{18}$$

with the energy distribution in front of the specimen. Neglecting the broadening of the Maxwellian energy distribution by the Boersch effect, this distribution is given by

$$\int_o^{\infty} n(E_1)dE_1 = \int_o^{\infty} bE_1^{\frac{1}{2}} \exp(-E_1/kT)dE_1 \quad (b = \text{constant}) \tag{19}$$

The integrand of Eq. (19) is plotted in Fig. 12 (curve 1). The hatched area covered by

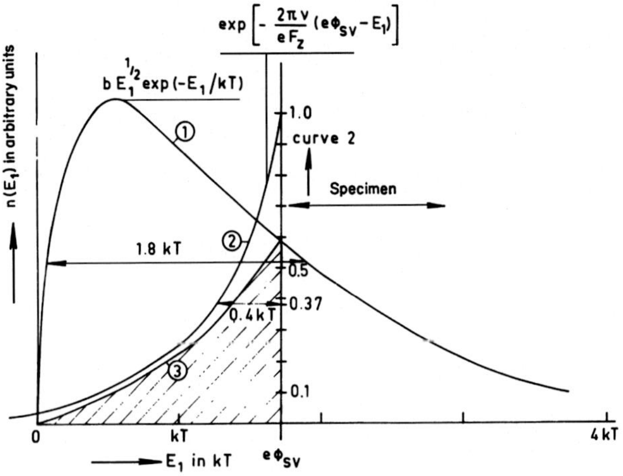

Figure 12 Effective contribution of electrons out of the Maxwellian energy distribution to the formation of image contrast.

$$\int_{o}^{e\phi_{SV}} \exp(-2\pi\nu z)n(E_1)dE_1$$
$$= \int_{o}^{e\phi_{SV}} \exp\left[-\frac{2\pi\nu}{eF_z}(e\phi_{SV}-E_1)\right] bE_1^{\frac{1}{2}} \exp(-E_1/kT)dE_1 \qquad (20)$$

represents the effective contribution of electrons out of the Maxwellian energy distribution to the formation of image contrast. The remaining part only gives a continuous background illumination in the final image. Curve 3 stands for the integrand of Eq. (20). The choice of $e\phi_{SV}$ as the upper limit in Eq. (20) involves the assumption that electrons reaching the specimen surface are ignored or in a more pessimistic view only contribute to the background illumination.

Introduction of a visibility criterion for the image contrast leads to a value of ν in Eq. (18) for which just enough lateral contrast is obtained.

Assuming 20% contrast yields a value for ν of

$$\frac{2\pi\nu}{eF_z}(e\phi_{SV}-E_1) = 1 \quad \text{with } (e\phi_{SV}-E_1) = 0.4kT.$$

The lateral resolving power (along the specimen) is then approximated by

$$\nu^{-1} = \frac{0.8\pi kT}{eF_z} = 70 \text{ nm} \quad (T = 2800 \text{ K}) \qquad (21)$$

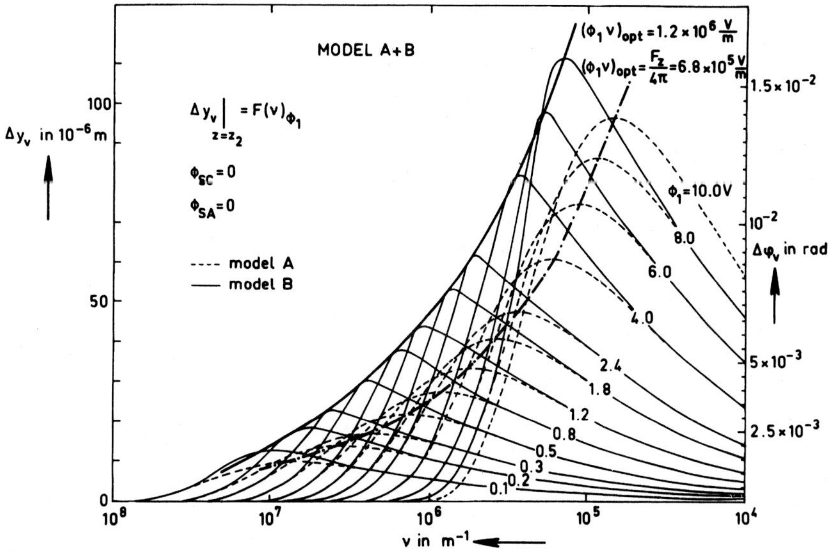

Figure 13 Modulation functions Δy_V and $\Delta\varphi_V$ plotted against ν (model A and B).

From Fig. 12 it is evident that already a rough filtering of the illuminating beam leads to a substantial improvement in contrast and therefore in lateral resolving power.

Concerning the lateral resolving power a point resolution of 80... 100 nm has been reached in Fig. 27B.

For the axial resolving power (step perturbations) the reader is referred to the appendix (Appendix B).

2.4 Model B (Solution of the Exact Equations of Motion 2.1a and b)

2.4.1 Electrostatic Contrast Calculations with a Digital Computer

By means of a digital computer a plot, similar to Fig. 5, for the maximum values of Δy_ν and $\Delta\varphi_\nu$ depending on ν and ϕ_1 as curve parameters, is presented in Fig. 13.

The condition $\sin 2\pi\nu y = 1$ of model A is not tenable for model B because it does not coincide with the maximum values of Δy_ν and $\Delta\varphi_\nu$. For comparison the curves of model A (dashed lines) and model B (full lines) are both pictured in Fig. 13. In this computer program the two second order differential equations were split into four first order equations. Integration of these equations is performed with the Nordsieck procedure,

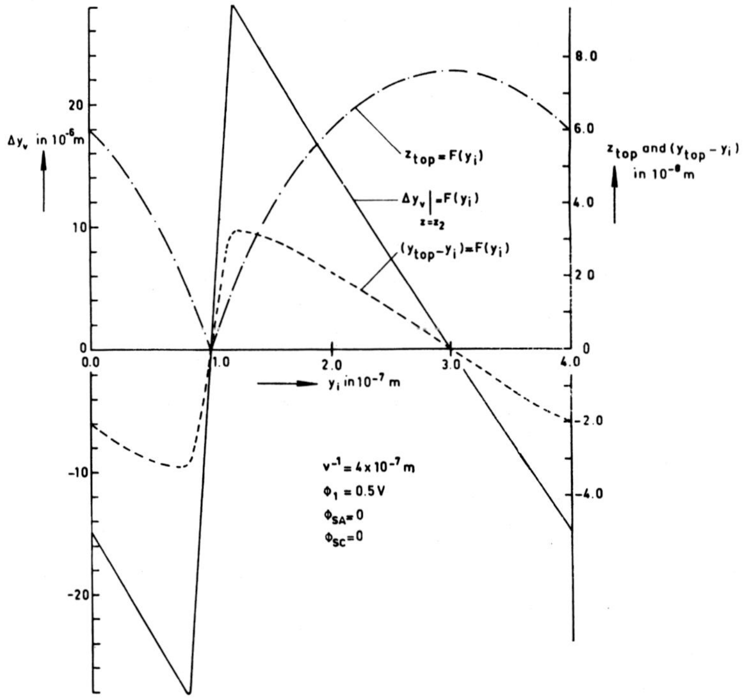

Figure 14 Δy_V and the coordinates y_{top} and z_{top} of the point of reversal plotted against y_i.

containing a variable steplength combined with a preset desired accuracy for each equidistant interval along the time axis.

As characteristic example of the computer results obtained Δy_v, $(y_{top} - y_i)$ and z_{top} are plotted (Fig. 14) against y_i the height of incidence, for a sinusoidal perturbation with $\nu^{-1} = 4 \times 10^{-7}$ m and $\phi_1 = 0.5$ V.

The coordinates z_{top} and y_{top} represent the point of reversal. For $y_i = 10^{-7}$ m, thus at the positive maximum of the perturbation, the electrons just reach the specimen surface, whereas for $y_i = 3 \times 10^{-7}$ m, z_{top} possesses its maximum.

For both values of y_i $(y_{top} - y_i) = 0$, thus the electron trajectories are straight lines.

Considering Figs. 13 and 14 the following conclusions for model B emerge:

1. For small values of Δy_v and $\Delta \varphi_v$, corresponding to a low contrast, both groups of curves match accurately as expected. Comparison of the

curves $\Delta y_\nu = F(\phi_1 \nu)\phi_1$ of model A (Fig. 6) with those of model B leads to a matching within 2% of A and B provided that

$$\frac{\phi_1 \nu}{F_z} < 1.2 \times 10^{-2} \quad \text{and} \quad \frac{\phi_1 \nu}{F_z} > 6 \times 10^{-1} \quad \left(F_z = 8.57 \times 10^6 \text{ V/m}\right) \tag{22}$$

Unfortunately, in practice, sufficient electrostatic contrast is only obtainable for values of $\frac{\phi_1 \nu}{F_z}$ between 1.2×10^{-2} and 6×10^{-1}. This means that the effect of $\frac{\partial \phi(y,z)}{\partial z}$ must be taken into account. For values of $\frac{\phi_1 \nu}{F_z}$ which fulfill condition (22), model B shows the linear behavior of model A.

In order to make model A applicable in practice, the circular contrast aperture D should be at least smaller than 10 μm. This value for D, however, is not feasible owing to the centering problems then arising. Since $\frac{\partial \phi(y,z)}{\partial z}$ has the same sign as F_z near positively charged spots, it is comprehensible that the maximum values of Δy_ν and $\Delta \varphi_\nu$ in model B exceed those of model A.

2. Like model A, the maxima of all B curves coincide with a line $\frac{(\phi_1 \nu)_{opt}}{F_z} = 1.4 \times 10^{-1}$.

For this value Δy_ν and $\Delta \varphi_\nu$ show almost a sawtooth shape instead of sinusoidal (see Fig. 14). A cosine perturbation thus causes a modulation effect consisting of a spectrum of sines, approximated by a constant $\times \sum_{n=1}^{\infty} n^{-1} \sin 2\pi n \nu y$.

The increasing damping effect (always present for electrostatic contrast) on higher order harmonics, both for model A and B, is demonstrated in Fig. 15, where Δy_ν is plotted against y_i for the following perturbations:

1. sine,
2. triangular,
3. trapezium,
4. square wave.

Except for differences in the maximum Δy_ν value, the shapes of the curves resemble each other closely. This indicates that although the modulation is by no means sinusoidal for the chosen values of ν and ϕ_1 almost no information about the accurate shape of an electrostatic perturbation can be obtained from the current density distribution in the final image.

2.4.2 Topographic Contrast

Concerning topographic contrast an exact solution of Eqs. (1a) and (1b) requires a relaxation of the electrostatic field near the specimen. Since these relaxation procedures lead to a fairly long computer time further calculations are omitted.

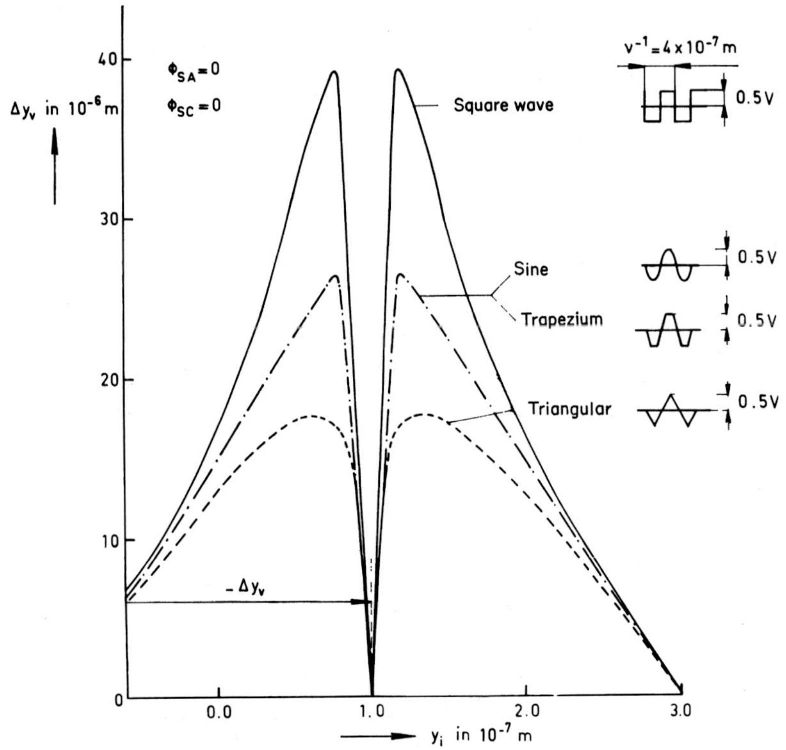

Figure 15 Modulation function Δy_V plotted against y_i for different wave forms.

2.4.3 Electrostatic Contrast Calculations with an Analog Computer

As a check on the digital computer results and to have a method which provides directly plotted results, Eqs. (1a) and (1b) were also solved with an analog computer. The functions $\frac{\partial\phi(y,z)}{\partial z}$ and $\frac{\partial\phi(y,z)}{\partial y}$ were partially generated with a combined sine and cosine potentiometer. The exponential damping factor in these derivatives was introduced into the machine with a ten point diode function generator. An increase of the interaction between the specimen perturbation and the reversing electrons is reached by letting the electrons enter the retarding field at $z = 4(2\pi\nu)^{-1}$ instead of $z = z_2$. The error involved, owing to $\phi\{y, 4(2\pi\nu)^{-1}\} \neq 0$, is $\exp(-2\pi\nu z) = 1.8 \times 10^{-2}$.

Addition of logic circuitry provided automatic plotting of the electron trajectories for given values of ϕ_o, ϕ_1, ν, δy_i (steplength along the y_i-axis), and α, the angle of illumination.

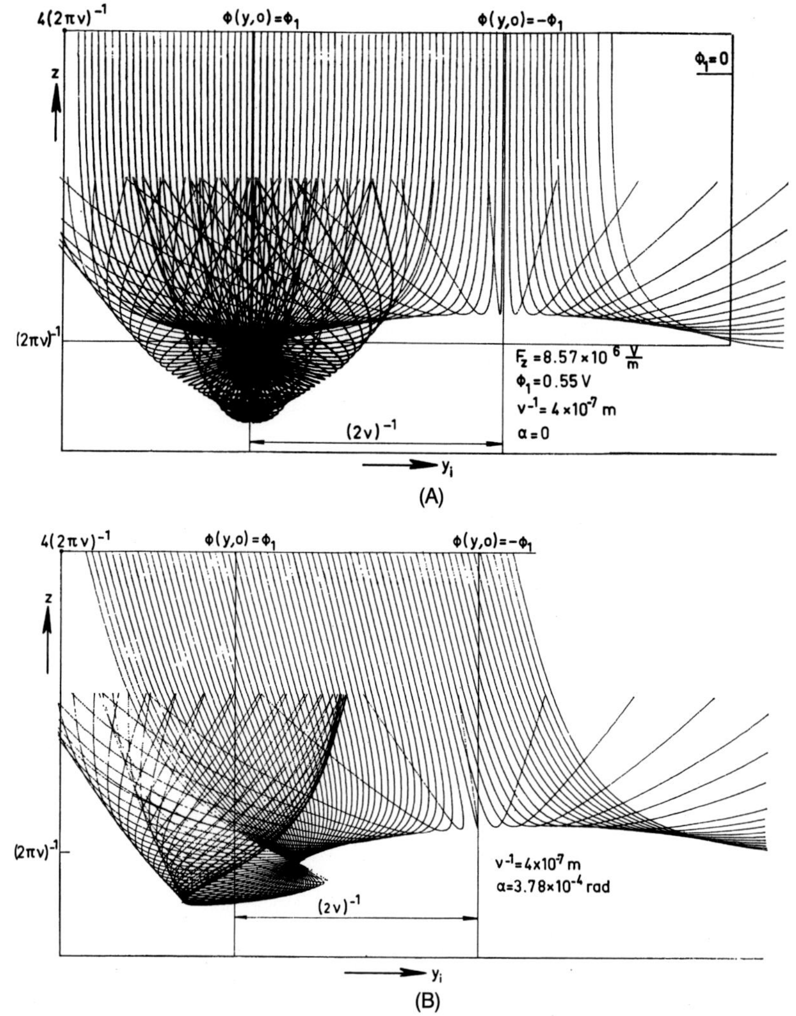

Figure 16 (A) Electron trajectories (model B) for $\alpha = 0$. (B) Electron trajectories (model B) for $\alpha = 3.78 \times 10^{-4}$ rad. (C) Electron trajectories (model B) for $\alpha = 7.56 \times 10^{-4}$ rad.

For three different angles α the electron trajectories are plotted against y_i in Figs. 16A, B and C with

$$\phi_o = 30 \text{ kV} \qquad \delta y_i = (80\nu)^{-1}$$

$$\phi_1 = 0.55 \text{ V} \qquad \nu^{-1} = 4 \times 10^{-7} \text{ m}$$

$$F_z = 8.57 \times 10^6 \text{ V/m} \qquad \phi_{SA} = \phi_{SC} = 0$$

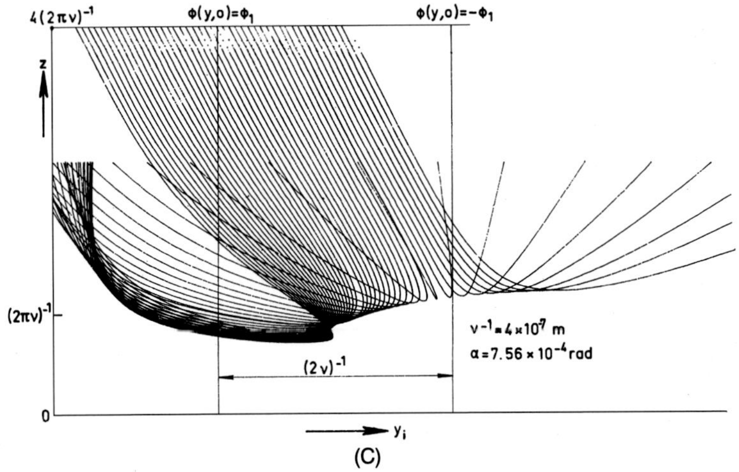

Figure 16 *(continued)*

The positively charged part of the sine wave clearly acts as a concave mirror. When the focus of this "micro mirror" is imaged it appears as a white line (corresponding to the stars in the actual pictures) in the final image. The negatively charged part is then related to a black band (bubbles in the actual pictures).

From the measured Δy_ν values relative current density distributions $\frac{j(y)}{j_o}$, referred back to the specimen, were plotted. As example $\frac{j(y)}{j_o}$ is presented for $\alpha = 0$ (dashed curve in Fig. 17). For this curve the angular aperture of the illuminating pencils is assumed to be infinitely small ($\beta \approx 0$). The filtering characteristic f_c, used is

$$f_c = 1 \quad \text{for } 0 \le G_1|\Delta y_\nu| \le D/2$$
$$f_c = 0 \quad \text{for } G_1|\Delta y_\nu| > D/2 \text{ with } D = 100\ \mu\text{m}$$

Since $\beta \approx 0$, the cross-over of the illuminating electron beam is small compared with D.

A remarkable and unexpected change in $\frac{j(y)}{j_o}$ occurs for a finite value of β or the related cross-over in the aperture plane. For a semi-angular aperture of only $\beta = 10^{-3}$ rad, $\frac{j(y)}{j_o}$ (full line in Fig. 17) has been plotted against y_i, extending for a full period $\nu^{-1} = 4 \times 10^{-7}$ m across the specimen surface. This curve has been constructed graphically by adding the relative current density distributions with $\beta = 0$ for six different but equidistant values of α. The cross-over size of the illuminating beam corresponding to

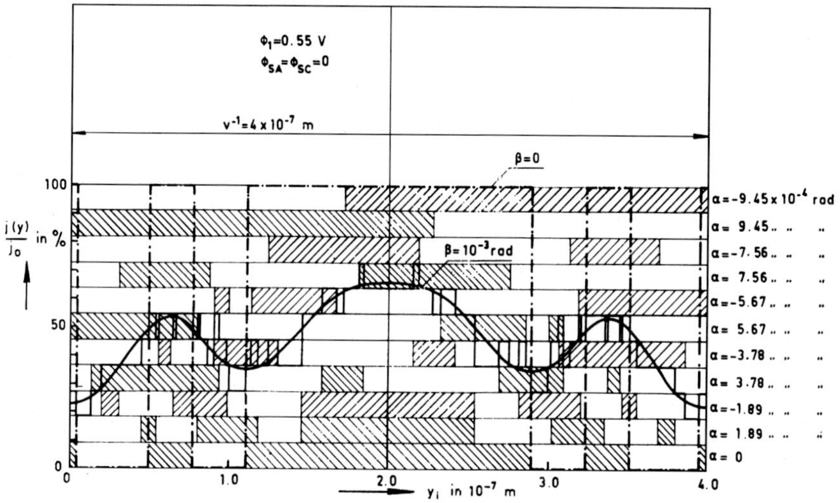

Figure 17 Relative current density distribution for $\beta = 0$ and $\beta = 10^{-3}$ rad.

$\beta = 10^{-3}$ rad amounts to 20 μm (focal length of the combination objective and electrostatic lens is 10 mm) and is therefore still small compared to $D = 100$ μm. Yet the influence of the cross-over size turns out to be tremendous.

The current density distribution for $\beta = 10^{-3}$ rad shows, although not exactly, the double frequency of the sinusoidal perturbation as a result of the double rectifying effect of the contrast aperture. The nonlinear behavior of model B for an electrostatic perturbation with $\phi_1 = 0.55$ V and $\nu^{-1} = 4 \times 10^{-7}$ m gives rise to the occurrence of rectified higher order harmonics in $\frac{j(y)}{j_o}$.

2.5 Model C

Although the application of wave mechanics to a "macroscopic" problem such as the formation of contrast is not expected to lead to other results than the classical calculations, it is performed in extension of the calculations of Wiskott (1956).

This model uses the time independent Schrödinger equation (2)

$$-\frac{\hbar^2}{2m}\left(\frac{\partial^2 u}{\partial y^2} + \frac{\partial^2 u}{\partial z^2}\right) + Vu = Eu \tag{2}$$

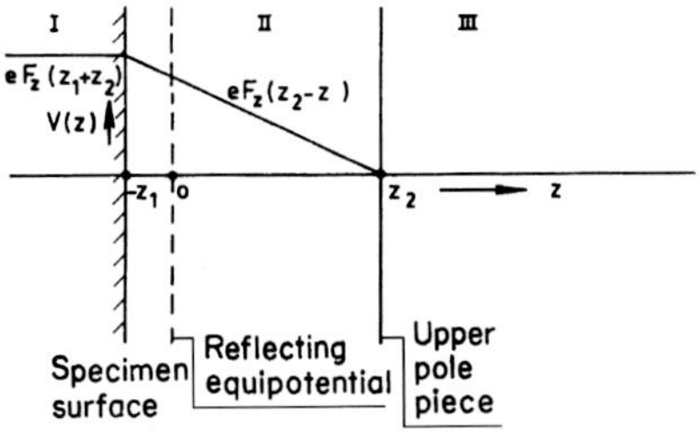

Figure 18 Potential energy for regions I, II, and III.

where

$u = u(y, z)$, the wave function
$V = V(z)$, the potential energy
$E =$ constant, the kinetic energy of the incident beam
$\hbar = \frac{h}{2\pi}$, h is Planck's constant.

In contrast with the calculations of Wiskott three regions are distinguished (see Fig. 18).

I. $z \le -z_1$, where $V = eF_z(z + z_2)$

$$F_z = \frac{E}{ez_2} = \frac{\phi_o}{z_2} = 8.57 \times 10^6 \text{ V/m}$$

II. $-z_1 < z < z_2$, where $V = eF_z(z_2 - z)$
III. $z \ge z_2$, where $V = 0$.

The separation into these regions matches the experimental set-up more accurately. Furthermore the retarding potential ϕ_o, although large with respect to must have a limited value.

The physical specimen surface is located at $z = -z_1$, whereas electron reflection occurs from the equipotential at $z = 0$. This configuration has been chosen here, contrary to model A, to include directly the effect of the specimen voltage ϕ_{SV}, which is related to z_1. Similar to model A, $|z_1| \ll z_2$ and $z_1 \to \frac{\phi_1}{F_z}$ provide maximum interaction of the specimen perturbation on the reversing electron beam. This wave mechanical calculation, performed in the yoz-plane, deals firstly with the unperturbed solution of Eq. (2) for regions I, II, and III.

The complete solution, a continuous wave function extending over the entire yoz-plane, is found by matching regions I, II, and III. This is achieved by fulfilling the boundary conditions at $z = -z_1$ and $z = z_2$

$$u_{u_I}|_{z=-z_1} = u_{u_{II}}|_{z=-z_1} \quad \text{and} \quad \left.\frac{du_{u_I}}{dz}\right|_{z=-z_1} = \left.\frac{du_{u_{II}}}{dz}\right|_{z=-z_1}$$
$$u_{u_{II}}|_{z=z_2} = u_{u_{III}}|_{z=z_2} \quad \text{and} \quad \left.\frac{du_{u_{II}}}{dz}\right|_{z=z_2} = \left.\frac{du_{u_{III}}}{dz}\right|_{z=z_2} \tag{23}$$

$u_{u_I}(z)$, $u_{u_{II}}(z)$, and $u_{u_{III}}(z)$ represent the unperturbed solutions corresponding to the regions mentioned above.

Since the perturbations on the specimen, in terms of voltage, are assumed to be small with respect to the retarding potential ϕ_o, the usual perturbation calculus can be applied. This allows for the determination of the perturbed wave function $u(z)$ from the unperturbed wave function $u_u(z)$. The calculations are simplified by neglecting second and higher order effects.

1. Determination of the unperturbed wave functions.

Region I: Eq. (2) changes into

$$-\frac{\hbar^2}{2m}\frac{d^2u_{u_I}}{dz^2} + u_{u_I}eF_z(z_1 + z_2) = eF_z z_2 u_{u_I}$$

$$u_{u_I} = C_1 \exp\left[\gamma\left(\frac{z_1}{z_2}\right)^{\frac{1}{2}} z\right] + D_1 \exp\left[-\gamma\left(\frac{z_1}{z_2}\right)^{\frac{1}{2}} z\right]$$

where

$$\gamma^2 = \left(\frac{mv_z}{\hbar}\right)^2 = \left(\frac{2\pi}{\lambda}\right)^2$$

For 30 kV electrons $\lambda = 7.07 \times 10^{-12}$ m and $\gamma = 8.88 \times 10^{11}$ m^{-1}. Obviously the second exponential term in u_{u_I} goes to infinity for large negative values of z, so $D_1 = 0$.

$$u_{u_I} = C_1 \exp(\gamma_1 z) \tag{24}$$

where

$$\gamma_1 = \gamma\left(\frac{z_1}{z_2}\right)^{\frac{1}{2}} = 1.50 \times 10^{13} z_1^{\frac{1}{2}} \text{ m}$$

$$0 \le z_1 < 10^{-7} \text{ m} \quad \text{and} \quad z_2 = 3.5 \times 10^{-3} \text{ m}$$

Region II: Eq. (2) changes into

$$-\frac{\hbar^2}{2m}\frac{d^2 u_{u_{II}}}{dz^2} + u_{u_{II}} eF_z(z_2 - z) = eF_z z_2 u_{u_{II}}$$

Substitution of $z = \zeta l$ yields

$$\frac{d^2 u_{u_{II}}}{d\zeta^2} + u_{u_{II}}\zeta = 0, \quad \text{where } l^3 = \frac{\hbar^2}{2meF_z} = \frac{z_2}{\gamma} = 1.64 \times 10^{-9}\ \text{m}$$

$$u_{u_{II}} = C_1 Ai(-z\gamma_2) + D_2 Bi(-z\gamma_2) \tag{25}$$

where

$$\gamma_2 = l^{-1} = 6.08 \times 10^8\ \text{m}^{-1}$$

$Ai(z)$ and $Bi(z)$ are Airy functions.[2]

Region III: Eq. (2) changes into

$$-\frac{\hbar^2}{2m}\frac{d^2 u_{u_{III}}}{dz^2} - eF_z z_2 u_{u_{III}} = 0$$

$$u_{u_{III}} = C_3 \exp(i\gamma z) + D_3 \exp(-i\gamma z)$$

Since the intensity of the illuminating beam is known, $C_3 = 1$

$$u^n_{III} = \exp(i\gamma z) - D_3 \exp(-i\gamma z) \tag{26}$$

2. On the basis of the boundary conditions, Eq. (23), the constants C_1, C_2, D_2, and D_3 are defined.

$$C_1 = -\frac{2i\gamma \exp(i\gamma z_2)\exp(\gamma_1 z_1)}{\pi\gamma_2(x_4x_1 - x_2x_3)}$$

$$C_2 = -\frac{2i\gamma x_2 \exp(i\gamma z_2)}{\gamma_2(x_4x_1 - x_2x_3)}$$

$$D_2 = \frac{2i\gamma x_1 \exp(i\gamma z_2)}{\gamma_2(x_4x_1 - x_2x_3)}$$

$$D_3 = -\exp(2i\gamma z_2)\frac{x_4x_1 - x_2\bar{x}_3}{x_4x_1 - x_2x_3}$$

[2] See Abramowitz and Stegun (1965).

where

$$\begin{aligned}
x_1 &= \frac{\gamma_1}{\gamma_2} Ai(z_1\gamma_2) + Ai'(z_1\gamma_2) \\
x_2 &= \frac{\gamma_1}{\gamma_2} Bi(z_1\gamma_2) + Bi'(z_1\gamma_2) \\
x_3 &= i\frac{\gamma}{\gamma_2} Ai(-z_2\gamma_2) - Ai'(-z_2\gamma_2) \\
x_4 &= i\frac{\gamma}{\gamma_2} Bi(-z_2\gamma_2) - Bi'(-z_2\gamma_2)
\end{aligned} \tag{27}$$

The bars indicate complex conjugate quantities.

Since $|D_3| = 1$, the illuminating beam is entirely reflected. As a result of $|D_3| = 1$ it follows that $C_1 = 0$, which is rather obvious because the chosen $V(z)$ distribution does not allow tunnel effects. Calculation of the current density $j_{z_{III}}$ of the unperturbed beam for region III provides an additional check.

$$\begin{aligned}
j_{z_{III}} &= \varrho_{III} v_z \ (\varrho = \text{electron density}) \\
j_{z_{III}} &= \frac{\hbar}{2im}\left[\bar{u}_{u_{III}} \frac{du_{u_{III}}}{dz} - u_{u_{III}} \frac{d\bar{u}_{u_{III}}}{dz}\right] \\
&= \frac{\hbar}{2im}\left(2 - 2D_3^2\right) i\gamma = v_z\left(1 - D_3^2\right)
\end{aligned}$$

For $|D_3| = 1 \rightarrow j_{z_{III}} = 0$, the incident beam is fully reflected.

Solution $u_{u_I} = C_1 \exp(\gamma_1 z)$ represents a stationary electron cloud for $z \leq 0$.

3. Superimposed on the equipotential at $z = -z_1$ is a sinusoidal perturbation

$$\phi_t(y, -z_1) = \phi_1 \cos 2\pi\nu y = \frac{\phi_1}{2}\left[\exp(2\pi i\nu y) + \exp(-2\pi i\nu y)\right]$$

$$\begin{aligned}
\phi_1 &= \text{perturbation amplitude in volts} \\
\nu &= \text{spatial specimen frequency}
\end{aligned}$$

Considering only the first exponential term $\frac{\phi_1}{2}\exp(2\pi i\nu y)$ in the following calculation, then

$$\phi(y, -z_1) = \phi_1 \exp(2\pi i\nu y)$$

According to Eq. (4)

$$\phi(y, z) = \phi_1 \exp(2\pi i\nu y)\exp(-|z + z_1|\nu) \tag{28}$$

With reference to the superimposed perturbing field $\phi(y, z)$ Eq. (2) for region II changes into

$$-\frac{\hbar^2}{2m}\left(\frac{\partial^2 u_{II}}{\partial y^2}+\frac{\partial^2 u_{II}}{\partial z^2}\right)-eF_z z u_{II}$$
$$=-u_{II}\eta\phi_1 \exp(2\pi i\nu y)\exp(-|z+z_1|\nu) \tag{29}$$

$u_{II} = u_{II}(y, z)$ represents the perturbed wave function for region II. It appears allowable to assume that the general perturbed wave function $u(z)$ consists of the unperturbed electron beam u_u, diminished with ηu_2, and a transverse beam ηu_1 modulated with the same periodicity ν as the specimen perturbation.

$$u = u_u + \eta u_1 \exp(2\pi i\nu y) - \eta u_2 \tag{30}$$

For instance for region II

$$u_{II} = u_{u_{II}} + \eta u_{1_{II}} \exp(2\pi i\nu y) - \eta u_{2_{II}}$$

The axially directed beam ηu_2 stands for the decrease of u_u owing to the generated transverse beam $\eta u_1 \exp(2\pi i\nu y)$. Factor η is an arbitrary small factor, which is introduced to distinguish higher order effects. Since this calculation only intends to describe first order effects, terms with η^2 and higher powers of η are omitted. Inserting Eq. (30) into (29) and similar equations for regions I and III yields

Region I:

$$\underbrace{-\frac{d^2u_{u_I}}{dz^2}+\gamma^2\frac{z_1}{z_2}u_{u_I}}_{\overset{\|}{0}}+\eta\exp(2\pi i\nu y)\left(-\frac{d^2u_{u_I}}{dz^2}+\gamma_3^2 u_{1_I}\right)$$
$$+\eta\left(\frac{d^2u_{2_I}}{dz^2}-\gamma^2\frac{z_1}{z_2}u_{2_I}\right)=-u_{u_I}\eta\phi_1\frac{2m}{\hbar^2}\exp(2\pi i\nu y)\exp(-|z+z_1|\nu)$$

Multiplying the remaining part with $\frac{1}{\eta}\exp(-2\pi i\nu y)$ and integrating between

$$y_1 = -\frac{\pi}{\nu}$$

and

$$y_2 = \frac{\pi}{\nu}$$

results in

$$-\frac{d^2u_{1_I}}{dz^2}+\gamma_3^2u_{1_I}=f_1$$

where

$$\gamma_3^2=\gamma^2\frac{z_1}{z_2}-4\pi^2\nu^2$$

and

$$f_1=-u_{u_I}\phi_1\frac{2m}{\hbar^2}\exp(-|z+z_1|\nu)$$

Variation of constants provides

$$\begin{aligned}u_{1_I}=&\ C_{1_I}\exp(\gamma_3z)+D_{1_I}\exp(-\gamma_3z)\\&+\frac{1}{2\gamma_3}\left[\exp(\gamma_3z)\int_{-\infty}^{-z}\exp(-\gamma_3z)f_1\,dz\right.\\&\left.-\exp(-\gamma_3z)\int_{-\infty}^{-z}\exp(\gamma_3z)f_I\,dz\right]\end{aligned}\tag{31}$$

Again $D_{1_I}=0$ to avoid u_{1_I} going to infinity for large negative values of z.

Region II: a similar procedure leads to

$$\begin{aligned}u_{1_{II}}=&\ C_{1_{II}}Ai(-z\gamma_4)+D_{1_{II}}Bi(-z\gamma_4)\\&-\pi\left[Ai(-z\gamma_4)\int_{-z_1}^{z}Bi(-z\gamma_4)f_{II}\,dz\right.\\&\left.-Bi(-z\gamma_4)\int_{-z_1}^{z}Ai(-z\gamma_4)f_{II}\,dz\right]\end{aligned}\tag{32}$$

where

$$\gamma_4=\left(\frac{\gamma^2}{z_2}-4\pi^2\nu^2\right)^{\frac{1}{3}}$$

and

$$f_{II}=-u_{u_{II}}\phi_1\frac{2m}{\hbar^2}\exp(-|z+z_1|\nu)$$

Region III:

$$u_{1_{III}}=C_{1_{III}}\exp(i\gamma_5z)+D_{1_{III}}\exp(-i\gamma_5z)$$

$$+\frac{1}{2i\gamma_5}\left[\exp(i\gamma_5 z)\int_{z_1}^{z}\exp(-i\gamma_5 z)f_{III}\,dz\right.$$
$$\left.-\exp(-i\gamma_5 z)\int_{z_2}^{z}\exp(i\gamma_5 z)f_{III}\,dz\right]$$

where

$$\gamma_5=\left(\gamma^2-4\pi^2\nu^2\right)^{\frac{1}{2}}$$

and

$$f_{III}=-u_{u_{III}}\phi_1\frac{2m}{\hbar^2}\exp(-|z+z_1|\nu)$$

Since the perturbations at the specimen are assumed to be small $f_{III}=0$. The illumination of the specimen with an electron beam parallel to the z-axis makes $C_{1_{III}}$ representing skew illumination, equal to zero.

$$u_{1_{III}}=D_{1_{III}}\exp(-i\gamma_5 z) \tag{33}$$

Summarizing

$$\lambda=7.07\times10^{-12}\ \text{m}$$
$$\gamma=8.88\times10^{11}\ \text{m}^{-1}$$
$$\gamma_1=\gamma\left(\frac{z_1}{z_2}\right)^{\frac{1}{2}}=1.50\times10^{13}z_1^{\frac{1}{2}}\ \text{m}$$
$$\gamma_2=\left(\frac{\gamma^2}{z_2}\right)^{\frac{1}{3}}=6.08\times10^{8}\ \text{m}^{-1}$$
$$z_2\gamma_2=2.13\times10^{6}$$
$$0\le z_1<10^{-7}\ \text{m}\quad\text{and}\quad z_1\ge\frac{\phi_1}{F_z}=1.17\times10^{-7}\phi_1\ \text{m}$$
$$\frac{\gamma}{\gamma_2}=1.46\times10^{3}$$
$$\gamma_3=\left(\gamma^2\frac{z_1}{z_2}-4\pi^2\nu^2\right)^{\frac{1}{2}}$$
$$\gamma_4=\left(\frac{\gamma^2}{z_2}-4\pi^2\nu^2\right)^{\frac{1}{3}}$$
$$\gamma_5=\left(\gamma^2-4\pi^2\nu^2\right)\approx\gamma$$
$$z_2=3.5\times10^{-3}\ \text{m}$$

4. Similar boundary conditions as Eq. (23) make the values of the perturbed wave functions u_{1_I}, $u_{1_{II}}$, $u_{1_{III}}$ and its first derivatives in z again coincident at $z = -z_1$ and $z = z_2$. For $z = -z_1$ and $z = z_2$ it follows that

$$u_{1_I}|_{z=-z_1} = C_{1_I} \exp(-\gamma_3 z_1) + E_{1_I} \exp(-\gamma_3 z_1) + F_{1_I} \exp(\gamma_3 z_1)$$

where

$$E_{1_I} = \frac{1}{2\gamma_3} \int_{-\infty}^{-z_1} \exp(-\gamma_3 z) f_I dz$$

and

$$F_{1_I} = -\frac{1}{2\gamma_3} \int_{-\infty}^{-z_1} \exp(\gamma_3 z) f_I dz$$

$$u_{1_{II}}|_{z=-z_1} = C_{1_{II}} Ai(z_1 \gamma_4) + D_{1_{II}} Bi(z_1 \gamma_4)$$
$$u_{1_{II}}|_{z=z_2} = (C_{1_{II}} + E_{1_{II}}) Ai(-z_2 \gamma_4) + (D_{1_{II}} + F_{1_{II}}) Bi(-z_2 \gamma_4)$$

where

$$E_{1_{II}} = -\pi \int_{-z_1}^{z_2} Bi(-z\gamma_4 z) f_{II} dz$$

and

$$F_{1_{II}} = \pi \int_{-z_1}^{z_2} Ai(-z\gamma_4 z) f_{II} dz$$

and

$$u_{1_{III}}|_{z=z_2} = D_{1_{III}} \exp(-i\gamma_5 z_2)$$

The boundary conditions for u_{1_I}, $u_{1_{II}}$ and $u_{1_{III}}$ lead to expressions for the constants C_{1_I}, $C_{1_{II}}$, $D_{1_{II}}$ and $D_{1_{III}}$.

$$C_{1_I} = \left[C_{1_{II}} Ai(z_1 \gamma_4) + D_{1_{II}} Bi(z_1 \gamma_4)\right] \exp(\gamma_3 z_1) - E_{1_I} - F_{1_I} \exp(2\gamma_3 z_1)$$

$$C_{1_{II}} = \frac{1}{x_6 x_7 - x_5 x_8} \left[2\gamma_3 x_8 F_{1_I} \exp(-\gamma_3 z_1) + E_{1_{II}} x_6 x_7 + F_{1_{II}} x_6 x_8\right]$$

$$D_{1_{II}} = \frac{1}{x_6x_7 - x_5x_8}\left[2\gamma_3x_7F_{1_I}\exp(\gamma_3z_1) + E_{1_{II}}x_5x_7 + F_{1_{II}}x_5x_8\right]$$
$$D_{1_{III}} = \frac{2i}{\gamma_5}\exp(i\gamma_5z_2)\left[(C_{1_{II}} + E_{1_{II}})\bar{x}_7 + (D_{1_{II}} + F_{1_{II}})\bar{x}_8\right] \quad (34)$$

where

$$x_5 = \gamma_3Ai(z_1\gamma_4) + \gamma_4Ai'(z_1\gamma_4)$$
$$x_6 = \gamma_3Bi(z_1\gamma_4) + \gamma_4Bi'(z_1\gamma_4)$$
$$x_7 = i\gamma_5Ai(-z_2\gamma_4) - \gamma_4Ai'(-z_2\gamma_4)$$
$$x_8 = i\gamma_5Bi(-z_2\gamma_4) - \gamma_4Bi'(-z_2\gamma_4)$$

Thus the determined general perturbed wave function contains as most interesting part u_{III_r}, representing the reflected beam in region III. Including also the negative exponential term

$$\frac{\phi_1}{2}\exp(-2\pi i\nu y)$$

of the sinusoidal specimen perturbation $\phi_t(y, -z_1)$, it is found that

$$u_{III_r} = D_3\exp(-i\gamma z) + \tfrac{1}{2}D_{1_{III}}\exp(-i\gamma_5z)\left[\exp(2\pi i\nu y) + \exp(-2\pi i\nu y)\right]$$
$$u_{III_r} \approx \exp(-i\gamma z)\left\{D_3 + \tfrac{1}{2}D_{1_{III}}\left[\exp(2\pi i\nu y) + \exp(-2\pi i\nu y)\right]\right\} \quad (35)$$

In this equation $D_3\exp(-i\gamma z)$ represents the unperturbed reflected beam (zero order) and the remaining part both transversely modulated beams as first order diffracted beams.

The results obtained resemble closely the diffraction pattern of an optical grating illuminated with a parallel beam of coherent light of wave length λ_l. On the basis of Abbe's diffraction formula $\Theta_l = \lambda_l\nu$, applied for a sinusoidal grating with a periodicity of ν the "effective" electron wave length λ_e in the perturbed retarding field can be compared with λ_l.

The diffraction angle Θ_e for the transverse beam can be derived from Eq. (35).

$$\Theta_e = \frac{\bar{u}_m\frac{\partial u_m}{\partial y} - u_m\frac{\partial \bar{u}_m}{\partial y}}{\bar{u}_m\frac{\partial u_m}{\partial z} - u_m\frac{\partial \bar{u}_m}{\partial z}} \quad (36)$$

where

$$u_m = \tfrac{1}{2} D_{1_{III}} \exp(-i\gamma_5 z) \exp(2\pi i \nu y)$$
$$\Theta_e = \frac{2\pi\nu}{\gamma_5} \approx \frac{2\pi\nu}{\gamma} = \lambda_e \nu \tag{37}$$

Setting $\Theta_e = \Theta_l$ provides $\lambda_e = \lambda_l = \lambda$.

Conclusion

The retarding voltage with the superimposed specimen perturbation $\phi(y, -z_1)$ behaves in first order, regardless of the strongly increasing electron wave length near the specimen, as an optical grating with an identical spatial frequency and irradiated with "light" of wave length λ_l equal to the electron wave length of the incident beam.

This conclusion, which involves that the modulation effect of the perturbations in the electrostatic field only occur near the specimen surface, leads directly to a criterion for the best obtainable lateral resolving power on the basis of Heisenberg's uncertainty principle

$$\Delta p_y \Delta y_s \geq h.$$

In this expression Δp_y stands for the tangential impulse determined by the product of objective lens angular aperture α_A and the axially directed impulse mv_z. The uncertainty in the location in the y-direction of an electron near the specimen surface then amounts to Δy_s where

$$\Delta y_s \geq \frac{h}{\alpha_A m v_z} = \lambda \frac{2f}{D} \tag{38}$$

For 30 kV electrons, $D = 10^{-1}$ mm and $f = 10$ mm (focal length of combination objective and electrostatic lens); it follows that

$$\Delta y_s \geq 1.6 \text{ nm}.$$

This figure is valid for coherent illumination and the absence of instrumental shortcomings such as spherical aberration of the retarding field and lens defects. In the case of incoherent illumination Δy_s increases by roughly a factor of two.

When $D_{1_{III}}$ is worked out, information can be obtained whether the perturbed specimen can be considered as a phase specimen or an amplitude specimen. Since no absorption of electrons occurs, it is obvious to state

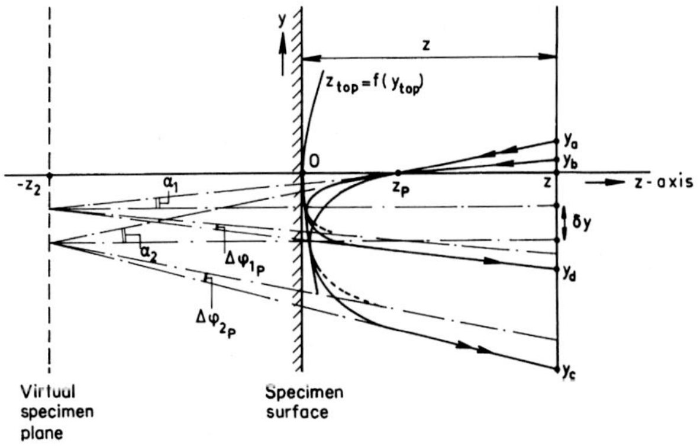

Figure 19 Electron trajectories in a mirror projection microscope. Calculation of the current density $j(y)_p$.

without further calculation of $D_{1_{III}}$ that the perturbed specimen must act as a phase specimen. A second argument in favor of the phase character of the specimen was found experimentally when a focused image was formed. Nearly no phase contrast appeared for values of D in the order of 500 μm and more. Only D values smaller than 200 μm gave sufficient contrast in the final image (see Fig. 26).

The current density distribution can be calculated from the double Fourier transform of the specimen perturbation.

2.6 Comparison with the Image Contrast for a Mirror Projection Microscope

An attempt will now be made to compare the contrast in a mirror microscope with focused images and in a mirror projection microscope. In the first instance only electrons incident at small angles with the z-axis are considered. For the mirror projection microscope an electron probe is formed at $z = z_p$ in front of the specimen. Although all electron trajectories through this electron probe are parabolas, having as envelope a paraboloid of revolution, electrons incident close to the z-axis can be considered to reflect from a flat equipotential plane at $z = z_1$, where $z_1 = \frac{\phi_{SV}}{F_z}$ and $z_p \gg z_1$. In first order approximation the angular deflection $\Delta\varphi_p$ for these electrons, as result of a sinusoidal specimen perturbation, follows from Eq. (10) belonging to model A. The index p refers to the projection method. The following calculations are only meant as a qualitative comparison between both types of contrast.

In the mirror projection microscope the formation of contrast emerges from local variations of the current density in a plane normal to the z-axis. It is shown that these variations are caused by the curvature of the perturbed specimen. In the projection microscope the plane $z = z_p$ is conjugate with the final screen. The relative current density distribution $\frac{j(y)_p}{j_p}$ in $z = z_p$ can be compared with $\frac{j(y)}{j_o}$. On the basis of Fig. 19 a relation between $\frac{j(y)_p}{j_p}$ and the specimen perturbation is derived. Both trajectories drawn represent electrons entering the retarding field at small angles α_1 and α_2.

Local perturbations cause angular deflections $\Delta\varphi_{1p}$ and $\Delta\varphi_{2p}$. Since all electrons inside the current tube, bounded by both trajectories, are reflected it follows that

$$j_p(y_a - y_b) = j(y)_p(y_d - y_c) \tag{39}$$

j_p is the current density of a reflected beam that is not modulated and $j(y)_p$ the current density of a modulated reflected beam; j_p and $j(y)_p$ are both measured in $z = z_p$. From Fig. 19 it follows that

$$\begin{aligned}
y_a &= y_a \\
y_b &= y_a - \delta y + 2z(\alpha_1 - \alpha_2) \\
y_c &= y_a - 4z\alpha_2 - 2z\Delta\varphi_{2_p} \\
y_d &= y_b - 4z\alpha_1 - 2z\Delta\varphi_{1_p} \\
\alpha &= \frac{v_y}{v_z} = \frac{y}{4z}; \quad \alpha_2 - \alpha_1 = \frac{\delta y}{4z}
\end{aligned}$$

Substitution of y_a, y_b, y_c and y_d into Eq. (39) leads to

$$z = z_p \rightarrow j_p = j(y)_p\left[1 - 4z_p\frac{\partial\Delta\varphi_p}{\partial y}\right]$$

Provided that $4z_p\frac{\partial\Delta\varphi_p}{\partial y} \ll 1$, which involves a slowly varying specimen perturbation, the relative current density distribution in $z = z_p$ is

$$\frac{j(y)_p}{j_p} = 1 + 4z_p\frac{\partial\Delta\varphi_p}{\partial y} \propto \frac{d^2\phi(y,0)}{dy^2} \tag{40}$$

The current density $j(y)_p$ is proportional to the curvature of the specimen for small values of $4z_p\frac{\partial\Delta\varphi_p}{\partial y}$.

Combination of Eq. (10) for $\Delta\varphi_p$ and Eq. (40) yields

$$\frac{j(y)_p}{j_p} = 1 + \frac{8\pi^2}{F_z}(2\nu^3 z_p)^{\frac{1}{2}}\phi_1 \cos 2\pi\nu y \exp(-2\pi\nu z_1) \tag{41}$$

An approximation of Eq. (19) is chosen for the calculation of the relative current density distribution $\frac{j(y)}{j_o}$ of the imaging mirror microscope

$$f_c = 1 - \frac{G_1|\Delta y|}{D}$$
$$\frac{j(y)}{j_o} = 1 - \frac{2\pi}{F_z}\frac{G_1}{D}(2\nu z_2)^{\frac{1}{2}}\phi_1 \sin 2\pi\nu y \exp(-2\pi\nu z_1) \tag{42}$$

For a 50% modulation in both relative current density distributions, which means

$$\frac{j(y)_p}{j_p} = \frac{j(y)}{j_o} = 0.5,$$

the crucial factors in the modulation terms of Eqs. (41) and (42) can be compared. It leads to the conclusion that the contrast for a mirror projection microscope is superior to that of a mirror microscope with focused images when

$$\nu > \frac{G_1}{4\pi D}\left(\frac{z_2}{z_p}\right)^{\frac{1}{2}} \tag{43}$$

Substitution of practical values

$$z_2 = 3.5 \text{ mm}; \quad z_p - 1 \text{ mm}; \quad G_1 = 1.5$$

leads to

$$\nu > \frac{2.2 \times 10^{-1}}{D} \tag{43a}$$

Application of a circular contrast aperture, for instance $D = 100$ μm, yields

$$\nu > 2.2 \times 10^3 \text{ m}^{-1}.$$

When the circular contrast aperture is replaced by a knife edge aperture, D stands for the cross-over of the illuminating beam. For magnifications at the final screen larger than 1500×, D is in the order of 1 μm so

$$\nu > 2.2 \times 10^5 \text{ m}^{-1} \tag{43b}$$

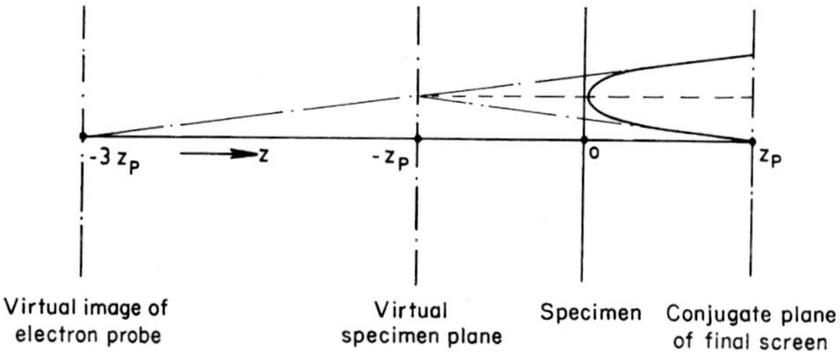

Figure 20 Calculation of the Fresnel diffraction at the specimen.

The size of the cross-over of the illuminating beam for magnifications larger than 1500× is mainly determined by the optical reduction of the electron source in the condenser system.

The conclusion that the contrast of a mirror projection microscope exceeds the contrast of a mirror microscope with focused images, in case Eq. (43a) or (43b) is fulfilled, is rather trivial. Since the mirror electron microscope deals with phase specimens a defocusing leads to an increase of contrast at the expense of resolving power. The through-focal series of photographs (Fig. 27) clearly shows this effect for the out of focus images. Near the in focus condition image contrast can be regained by either decreasing the diameter D of the contrast aperture or reduction of the illuminating cross-over for the knife edge technique.

Not considered for the projection method is the blur due to the out of focus Fresnel diffraction. This type of diffraction results from the interference of reflected electrons from different parts of the specimen. Since the retarding field macroscopically acts as a flat mirror, the specimen can be considered as being illuminated in transmission by the virtual electron probe at $3z_p$ behind the specimen (see Fig. 20). The distance d_F, calculated back at the specimen, between the geometrical projection and the maximum of the first fringe of the diffraction pattern amounts to

$$d_F|_{z=0} = (z_p\lambda_p)^{\frac{1}{2}}$$

where λ_p is the electron wave length near $z = z_p$.

The conclusion of Section 2.5 justifies the use of this equation as a criterion for the theoretical resolving power in a mirror projection microscope

$$d_F|_{z=0} = (z_p\lambda_p)^{\frac{1}{2}} = \left(z_p\frac{\lambda_e^2\phi_o}{F_z}\right)^{\frac{1}{4}} = \left(z_p z_2 \lambda_e^2\right)^{\frac{1}{4}} \tag{44}$$

For 30 kV electrons the wave length λ_e of the incident beam amounts to 7×10^{-12} m. For $F_z = 8.6 \times 10^6$ V/m and $z_p = 10^{-3}$ m

$$d_F|_{z=0} = 1.2 \times 10^{-7}\ \text{m} \quad \left(\nu_F = 8.3 \times 10^6\ \text{m}^{-1}\right).$$

Because of the fourth root, the value of d_F is hardly affected by the distance z_p of the electron probe in front of the specimen. z_p must be sufficiently large to obtain an adequate field of view. In view of this theoretical resolving power, a resolution of 10^{-7} m (along the specimen surface) with the mirror projection microscope claimed by Litton Systems Inc. (1968) seems rather doubtful and is certainly not shown.

For the off axis electrons in a mirror projection microscope, the z coordinate of the point of reversal moves away from the specimen surface and causes, in first order, an additional exponential damping on the contrast. This damping effect depends on the locus of $y_{top} = f(z_{top})$, where y_{top} and z_{top} represent the coordinates of the point of reversal.

The locus is calculated for an unperturbed specimen. Perturbations present only give rise to slight local deviations.

According to Eq. (17) (see Fig. 19)

$$z_{top} = \left(\frac{v_y}{v_z}\right)_p^2 z_p = \frac{y_{top}^2}{4z_p} = \frac{y_p^2}{16z_p} \tag{45}$$

$(\frac{v_y}{v_z})_p$ and y_p refer to the plane $z = z_p$.

This equation represents a parabola, having its apex at $z = 0$ or at $z = z_1$ for $\phi_{SV} \neq 0$ and a curvature $\varrho_c \approx \frac{1}{8z_p}$.

For small values of z_p, ϱ_c tends to infinity. It means that an electron probe is formed at the specimen surface. Although in a mirror projection microscope without beam separation a focused image of the specimen is then formed, the field of view is very limited. Scanning the electron probe across the specimen surface provides a focused image with sufficient field of view. A scanning mirror electron microscope has been described by Garrood and Nixon (1968). Use of large values of z_p nullifies ϱ_c, or the specimen is illuminated with an electron beam parallel to the z-axis.

The additional distance z_{top} from the reflecting equipotential ($z = z_1$) can be introduced into Eq. (8) by

$$dt = -\left(\frac{m}{2eF_z}\right)^{\frac{1}{2}} (z - z_1 - z_{top})^{\frac{1}{2}}\, dz$$

Neglecting again the additional modulation damping during traveling of the reversing electrons parallel to the specimen surface,

$$\Delta\varphi_p = \frac{\pi}{F_z}\left(\frac{2v}{z_2}\right)^{\frac{1}{2}}\phi_1 \sin 2\pi\nu y \exp\left\{-2\pi\nu\left(z_1 + \frac{y_p^2}{16z_p}\right)\right\}$$

Since

$$\frac{y_p}{8z_p} = \frac{1}{2}\left(\frac{v_y}{v_z}\right)_p \ll 1$$

$$\frac{j(y)_p}{j_p} \approx 1 + \frac{8\pi^2}{F_z}\left(2v^3 z_p\right)^{\frac{1}{2}}\phi_1 \cos 2\pi\nu y \exp\left\{-2\pi\nu\left(z_1 + \frac{y_p^2}{16z_p}\right)\right\}$$

This equation shows that the modulation factor $\frac{j(y)_p}{j_p}$ for a mirror projection microscope falls off exponentially for off axis areas of the specimen. This effect is observable in most mirror projection pictures (see Fig. 21A). When the specimen is biased slightly positive so that $-z_1 = \frac{y_p^2}{16z_p}$ the exponential damping effect decreases at the expense of electrons striking the specimen surface near the axis. Then the image quality near the axis deteriorates owing to electron scattering at the surface, whereas the off axis areas reveal more detail (see Fig. 21B).

2.7 Magnetic Contrast

The modulation effect of a magnetic perturbation on the approaching electrons is in first order approximation equal but oppositely directed to the modulation on the reflected electrons. This means that in the case of normal incidence on the reflecting equipotential plane practically no magnetic contrast can be obtained.

The Lorentz force $\overline{K}$ caused by the magnetic perturbation, shows that for skew incidence magnetic contrast can be obtained.

$$\begin{aligned}
&\overline{K} = -(v \times \overline{B})\\
&K_x = -(ev_z B_y + ev_y B_z)\\
&K_y = -(ev_x B_z + ev_z B_x)\\
&K_z = -(ev_y B_x + ev_x B_y)\\
&\overline{B} = \text{magnetic induction}
\end{aligned} \tag{46}$$

K_z can be neglected because this component only modulates the depth of penetration into the retarding field. Since v_z changes sign at the point of

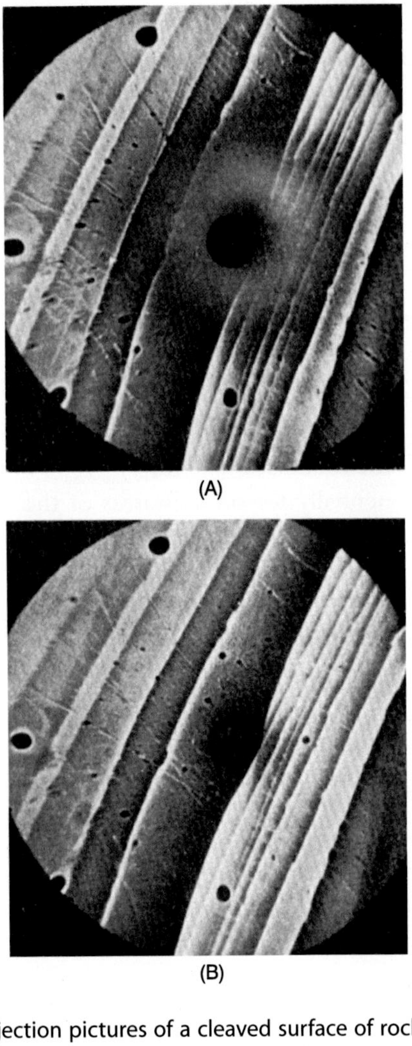

Figure 21 Mirror projection pictures of a cleaved surface of rock salt covered with an evaporated gold layer. Magnification about 100×. (These pictures were made by M.E. Barnett and W.C. Nixon, and are published with their kind permission.) (A) The specimen is negatively biased with −0.2 V. (B) The specimen is positively biased with +1.8 V.

reversal all deflection terms with v_z compensate one another if model A, with no lateral displacement during reversal, is used.

The remaining effect $K_r = (K_x^2 + K_y^2)^{\frac{1}{2}} = ev_t B_z$ shows that magnetic contrast only originates (in model A) from the interaction of the normal component B_z of the magnetic perturbing field and the tangential electron velocity v_t. Therefore only skew illumination reveals magnetic contrast.

In projection images magnetic contrast increases with the distance from the axis. This distance is proportional to v_t. This effect is known from Kranz and Bialas (1961). It is expected that the calculations according to model A are an even coarser approximation than those for electrostatic contrast.

According to a suggestion of Bethge sufficient magnetic contrast for in focus images can be achieved by converting the magnetic perturbing effect into an electrostatic perturbation. Therefore a magnetic perturbed specimen is covered with an insulating film and a layer which possesses a large Hall effect.

The emerging potential distribution, which occurs for current flow through this Hall layer, can be imaged.

2.8 Concise List of Symbols Used in Section 2

x, y, z	Cartesian coordinates; x and y coincide with the specimen surface; z along the electron optical axis
v_y, v_z	electron velocity in the y- and z-direction
z_1	coordinate of the reflecting equipotential plane
z_2	coordinate of the upper pole piece of the objective lens
z_p	coordinate of the electron probe (projection method)
λ_2	electron wavelength at $z = z_2$
v_2	electron velocity at $z = z_2$
ϕ_o, F_z	retarding potential and retarding field strength
$\Phi(y, z)$	potential distribution between the perturbed specimen and upper pole piece of objective lens
$\phi(y, z)$	perturbation potential
ϕ_s	specimen voltage
ϕ_{SC}	contact potential between specimen and tungsten filament
ϕ_{SA}	additional negative specimen bias
ϕ_n	perturbation amplitude of the n-order harmonic
ν	spatial specimen frequency
$d(y, 0)$	topography of the specimen
y_i	height of incidence at $z = z_2$
y_r	intersection of reflected electron with $z = z_2$
y_{top}, z_{top}	coordinates of the point of reversal
$\Delta y_v, \Delta \varphi_v$	lateral displacement and angular deflection, both at $z = z_2$ for electrostatic contrast
$\Delta y_t, \Delta \varphi_t$	lateral displacement and angular deflection, both at $z = z_2$ for topographic contrast

$\Delta y_1, \Delta\varphi_1$	lateral displacement and angular deflection in the contrast aperture plane
D	diameter of contrast aperture
$D_o = 2R_o$	diameter of the upper pole piece of the objective lens
f_c	filtering characteristic
j_o	current density of the parallel illuminating beam (in the retarding field)
$j(y)$	modulated current density, referred to the specimen
j_p	current density of reflected beam, being not modulated, at $z = z_p$
$j(y)_p$	modulated current density of reflected beam at $z = z_p$
M_1	magnification between specimen and object plane of the projector lens
β	semi-angular aperture of the electron pencils at $z = z_2$
α	angle of illumination at $z = z_2$
$u = u(y, z)$	wave function
$V = V(z)$	potential energy in the retarding field
E	kinetic energy of the incident beam
E_1	arbitrary energy in the one-dimensional Maxwellian energy distribution
$u_{u_I}, u_{u_{II}}, u_{u_{III}}$	unperturbed wave functions for regions I, II, and III
u_I, u_{II}, u_{III}	perturbed wave functions for regions I, II, and III
η	arbitrary small factor
$\left(\frac{v_y}{v_z}\right)_p, y_p$	quantities measured in $z = z_p$

3. DESCRIPTION AND DESIGN OP A MIRROR ELECTRON MICROSCOPE WITH FOCUSED IMAGES

3.1 Description

Apart from the requirement of separated axes for the illuminating and imaging system, a continuously variable magnification of 250...4000× at the final fluorescent screen was desired. On the basis of Fig. 22 the instrument will be described by following the accelerated electron beam in the illuminating system toward the specimen surface (the mirror electrode) and after reflection toward the final fluorescent screen. Fig. 23 shows an overall picture of the microscope.

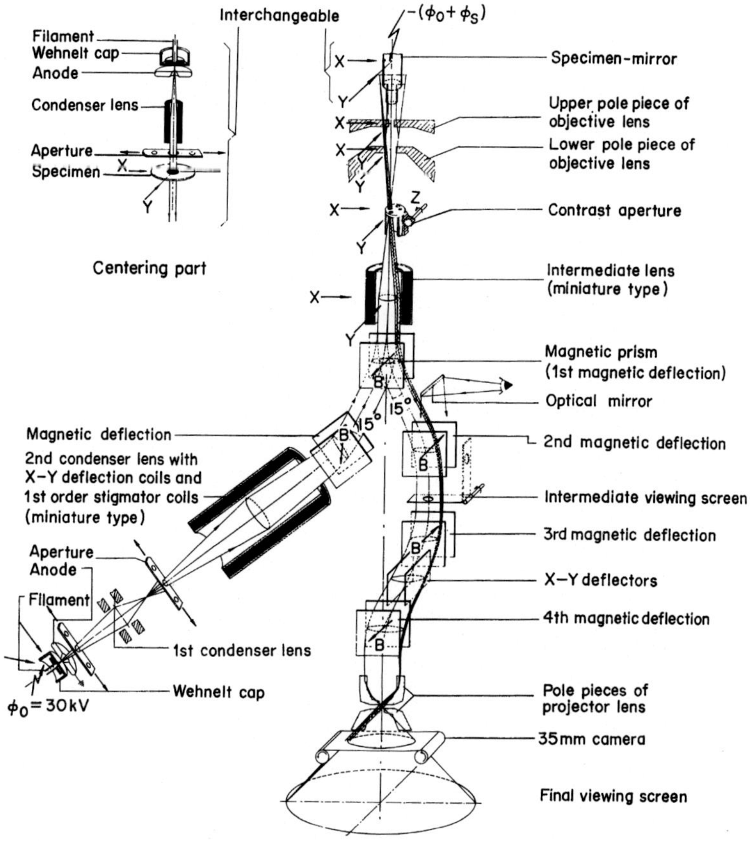

Figure 22 Ray diagram.

3.1.1 The Illuminating System

When a magnetic prism is used, it is favorable to minimize the deflection angle with regard to errors caused by the deflection field. These effects are proportional to the second and higher powers of the angle of deflection.

In order to create sufficient clearance for the miniaturized illuminating system, the angle between this system and the vertical main axis is fixed at 30°. The main axis represents the centerline of the specimen and the projector lens. The deflection angle of the magnetic prism is further reduced to 15° by mounting on top of the second condenser lens an additional deflector, which matches the axis of the illuminating system with the proper direction of incidence for the prism.

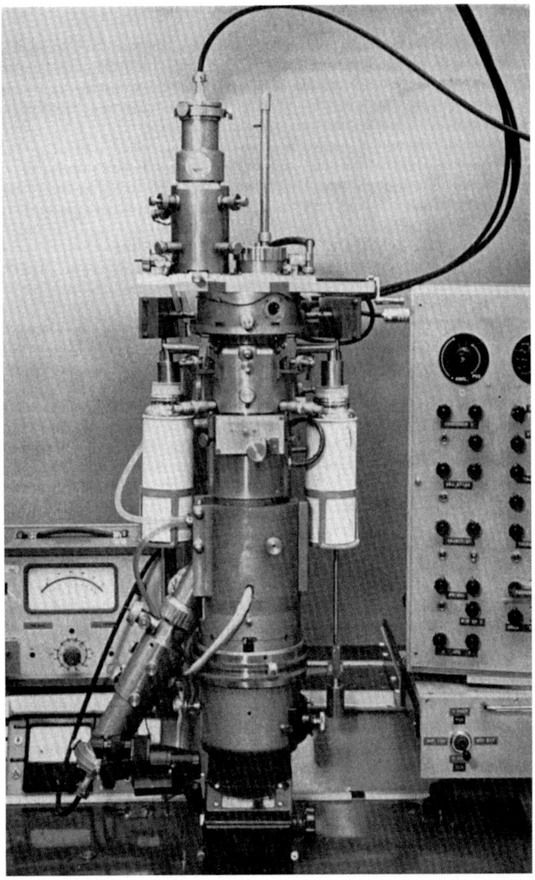

Figure 23 Mirror electron microscope for focused images.

A 30 kV electron beam is produced by a conventional triode electron gun. The first condenser lens with iron pole pieces demagnifies the electron source 10...40×. The second condenser lens, a miniature magnetic lens without iron circuit (Le Poole, 1964a), images the demagnified electron source through the injector-deflector, prism, and intermediate lens into the contrast aperture.

An adjustable holder for three apertures is mounted between the first and second condenser lens in order to obtain a fixed angular aperture of the illuminating electron beam.

Each deflector (the injector-deflector, the prism, and the later discussed additional "bridge-deflectors") consists of pairs of circular air coils. The inner sides of the coils are covered with thin sheets of transformer iron. The

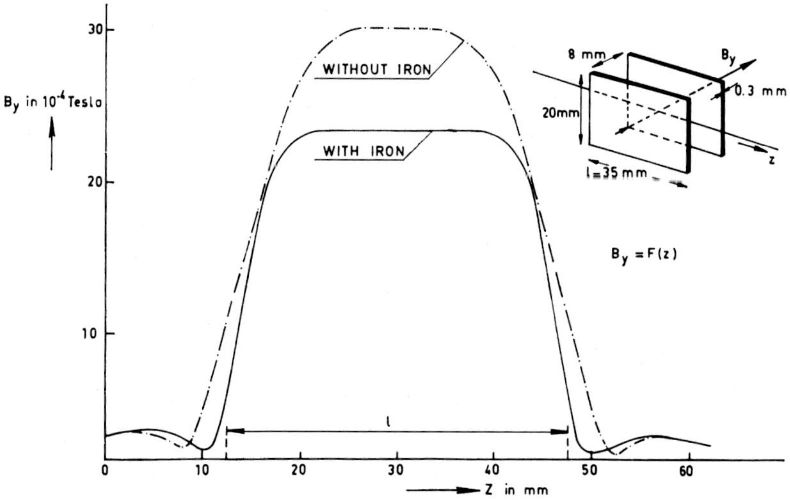

Figure 24 Transverse magnetic induction $B(y)$ of a deflector.

resulting magnetic field is quite homogeneous and has an almost rectangular boundary (Fig. 24).

Experiments revealed that, for electron beams with a diameter nearly equal to the separation of the iron sheets, the image distortion still remains within admissible limits. Without iron sheets the usable cross-section of the deflection field, with respect to image distortion, is too restricted for application in this instrument. The small amount of iron in the sheets leads to a negligible non-linearity.

The second condenser lens is provided with additional pairs of x- and y-deflectors for centering the lens and two quadrupoles for correcting astigmatism. Iron tubing screens the illuminating system against stray magnetic fields. Near the second condenser lens the iron tubing is connected with a rectangular iron plate, covering a hole in the vertical main column housing.

Care has been taken to separate the magnetic fluxes in the illuminating system from the fluxes in the main column. Especially, interaction of magnetic fluxes generated in different parts of the instrument gives rise to problems in the centering of the electron optics. In order to minimize this coupling effect, additional concentric iron cylinders are used near the joining of the illuminating system and the main column, and around the intermediate lens.

3.1.2 Imaging System

The imaging system consists of the objective lens with contrast aperture and the intermediate lens. The combination of objective, intermediate and projector lens allows a continuously variable magnification of 250…4000× at the final fluorescent screen.

The imaging system and the prism form a group of electron optical components which are passed both by the illuminating and the reflected beam. This feature sets high requirements on the centering accuracy of both lenses. The axes of both lenses should coincide perfectly with the main axis of the vertical column, because each residual inclination or decentering produces a transverse magnetic field which acts as a prism. In order to make the lens axes, in practice always inclined and decentered, coincident with the main axis, these lenses are centered by combining current reversing with pole piece centering, as proposed by Haine (1947).

Limitation of the field of view can be avoided by positioning the contrast aperture, as mentioned in Section 2, in the back focal plane of the combination objective and electrostatic lens. The negative lens action of the upper objective lens pole piece, which forms the earthed boundary of the retarding field, necessitates a slightly higher excitation of the objective lens to assure normal incidence onto the mirror plane. Since for changes in the magnification the objective lens excitation has to be varied, the contrast aperture is, apart from the x- and y-centering, also adjustable along the main axis (z-direction). To each setting of the objective lens current there corresponds an optimum z-position of the contrast aperture providing maximum field of view. Both the objective and the intermediate lens are provided with two crossed quadrupoles for correcting astigmatism.

In order to prevent magnetic flux interaction of the objective and the intermediate lens, the objective lens has been magnetically insulated from the surrounding iron tubing by positioning the iron circuit between two brass plates.

3.1.3 The Specimen Stage

The separation between the upper pole piece of the objective lens and the specimen plane amounts to 3.5 mm. For a retarding voltage of 30 kV the corresponding strength F_z of the electrostatic field measures 8.57×10^8 V/m. Although a higher F_z value would improve the image quality, this value proved to be a proper choice to minimize electrical breakdowns. Once an electrical breakdown occurs the specimen will usually be damaged by a spark-over or heavy ion bombardment (see Fig. 26D). The specimen carrier is a massive metal cylinder with a diameter of 10 mm.

The high sensitivity of this microscope for differences in topography requires for most applications a polished specimen surface. Therefore most of the specimens pictured in Section 4 consist of vacuum deposited layers onto an accurately polished glass disk, fitting into the specimen carrier.

The electrical conductivity of the specimen surface must be adequate to maintain the electrical potential. So when insulating materials have to be examined with this microscope, a conductive coating of the specimen is required. Since the retarded electrons possess rather low velocities near the specimen, the effective cross-section for ionization of the gas molecules present increases considerably. The positive ions thus generated bombard the negatively biased specimen and cause specimen damage. Moreover, a contamination layer of cracked organic molecules will be deposited onto the specimen surface.

This makes an ultrahigh vacuum (pressures lower than 10^{-6} $\mathrm{N\,m^{-2}}$) desirable. A further reduction of specimen contamination during observation is obtained by heating the specimen up to 50°C.

The specimen stage, consisting of two concentric stainless steel cylinders, of which the exterior cylinder is fixed and the interior one movable, is supported on a glass plate and insulated from earth potential. To minimize high electrical field concentrations the specimen surface is shielded (except for a central hole of 6 mm) with a rounded and polished stainless steel electrode.

In order to avoid an additional lens action due to this screening cap and to maximize the electrical field at the specimen surface, the specimen is mounted directly behind the electrode. The thickness of this electrode near the specimen surface measures 0.3 mm.

By means of three glass insulators the specimen can be translated within a circle of 6 mm in diameter.

Positioning of the specimen surface perpendicular to the main axis for normal mirror microscopy, or tilting the specimen with regard to the main axis over a controllable angle for dark field mirror microscopy, is performed by moving the stage across a sphere with its center at the intersection of the main axis and the specimen surface. Within mechanical tolerances this construction permits tilting of the specimen without translation.

The operation of this instrument is facilitated by an air lock system for changing specimens. Both the air lock system and a second electron gun, which is used for alignment of the microscope (see Section 3.1.7), are adjustable normal to the main axis. At will, either the air lock system or the second electron gun can be centered on the main axis.

Moreover, since the specimen stage protrudes into the cup-shaped upper pole piece of the objective lens, sufficient shielding near the specimen surface against stray magnetic fields is assured.

3.1.4 The Deflection Bridge and the Projector Lens with Camera

In previous mirror electron microscopes, equipped with a magnetic prism (Mayer, Rickett, & Stenemann, 1962; Bartz et al., 1956; Schwartze, 1967) the reflected electron beam (after passing the magnetic prism) is observed by a skew projection system.

In this microscope the reflected beam, after passing the prism, is made to coincide again with the main axis by means of three additional deflectors. The magnetic prism with the three following deflectors form the deflection bridge.

The advantages of making the reflected beam again coincident with the main axis are:

1. The effective deflection of the bridge is zero. Therefore this system shows an achromatic behavior for high voltage fluctuations. In addition, when the deflectors of the bridge are energized in series, correction against fluctuations in the series current is established. Without this compensation the required current stability for the prism amounts to a few parts per million. The use of the deflection bridge lowers the stability required for achieving identical quality in the final image at least by two orders of magnitude;
2. Except for the illuminating system, the main column can be erected vertically, which makes it easier to achieve the high requirements for mechanical stability;
3. A considerable facilitation for the alignment of the electron beam through the microscope is achieved (Section 3.1.7).

As an aid for alignment of the reflected beam through the bridge an intermediate fluorescent screen can be inserted between the second and third deflector. This screen can be observed through a glass window and an obliquely positioned glass mirror.

Thanks to the insensitivity of the deflection bridge for variations of the series current over a wide range (0.3…0.6 A) a current setting has been selected for which the astigmatism, which is a by-product of the deflection fields and results in different magnifications in perpendicular directions, is minimized. Residual astigmatism is corrected with a stigmator mounted

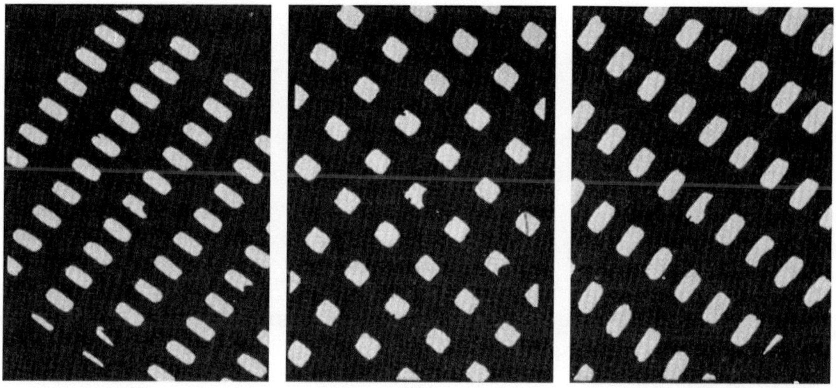

Figure 25 Images of a grid, illuminated in transmission, for different excitations (i_{defl}) of the deflector bridge. Magnification 120×. (A) $i_{defl} = 0.38$ A; (B) $i_{defl} = 0.46$ A (corrected); (C) $i_{defl} = 0.52$ A.

around the fourth deflector. Figs. 25A, B, and C clearly show the effect of the bridge excitation on the image distortion at the final screen.

Between the third and fourth deflector an additional set of x- and y-deflectors is mounted for precise alignment of the electron beam. The projector lens with camera housing and final fluorescent screen is of standard Philips design (EM 75) and described by Le Poole (1964b).

A continuously variable projector lens magnification (35...110×) is available by varying the pole piece separation. Image registration is performed either on 35 mm film, mounted directly below the projector pole pieces, or on a 60 × 60 mm^2 plate camera below the final fluorescent screen. A 7× binocular viewer permits accurate focusing.

3.1.5 Electrical Supplies

The high voltage generator produces a continuously variable accelerating voltage between 0 and 30 kV with an overall stability of better than 20 parts per million over a period of 15 min. The filaments in the electron guns are heated with a 1% stable d.c. current. Emission control is obtained by adjusting the Wehnelt resistor.

A variable d.c.-power supply, insulated against 30 kV, heats the specimen furnace. The specimen temperature can be measured with a thermocouple. The tungsten furnace coil is wound bifilarly to minimize stray magnetic fields near the specimen surface.

The stabilized voltage ϕ_s, variable between −12 V and +12 V, is superimposed on top of the retarding potential to prevent electrons from reaching the specimen surface and to correct for the contact potential ϕ_{SC}.

Electrical instabilities in the current supplies for the second condenser, objective, intermediate lens and all deflectors are stable within 5 parts per million over 15 min.

Most lens stigmators and x- and y-deflectors are energized in series with the lens currents. All electrical supplies are remote controlled in order to reduce stray magnetic fields in the vicinity of the column.

3.1.6 Vacuum System

In this microscope two oil diffusion pumps (baffled pumping speed about 100 l/s) provide a vacuum in the order of 5×10^{-3} N m^{-2} in the main column and 5×10^{-4} N m^{-2} near the specimen. Both pumps are backed by a mercury booster pump in combination with a rotary pump for roughing. The booster pump with additional vacuum vessel allows the switching off of the rotary pump after prevacuum pumping of the column.

An air lock system is incorporated for reloading the plate camera without breaking the vacuum in the main column and the specimen chamber.

3.1.7 Alignment Procedures

Owing to the large number of electron optical components, the alignment of the electron beam through this microscope should preferably be performed in steps. In this microscope the alignment is achieved on the basis of four procedures. The first procedure starts with the centering of the second electron gun, on top of the specimen stage, with respect to the projector lens.

The other lenses and deflectors are not energized, and the specimen stage is kept earthed. The second gun, centered on the projector lens, defines the main axis. Next, the axis of the objective lens is made coincident with this main axis by means of current reversing and pole piece centering.

After following a similar centering method for the intermediate lens, the contrast aperture is inserted in the electron beam.

The second procedure involves the electrical adjustment of the deflection bridge. Proper excitation of the bridge and the accessory stigmators provides a distortion free image on the final screen.

For the third procedure the second electron gun is earthed. The illuminating system then focuses the first electron source onto the filament tip of this gun. When the resulting current flow from the filament tip to earth potential only shows minor changes for current reversal of the objective and intermediate lens, the adjustment of the illuminating system is complete. The fourth and final alignment procedure involves biasing the specimen stage to the proper retarding voltage and positioning the specimen plane normal to the main axis.

All current settings of the critical electron optical components are quickly readable from a four decade digital voltmeter. This control feature reduces considerably the time required for accomplishing the four alignment procedures.

For a calculation of the electron optical parameters and the ray diagrams for both the real and the virtual imaging mode of the objective lens the reader is referred to the author's thesis (Bok, 1968).

4. RESULTS AND APPLICATIONS

4.1 Results

Since interpretation of mirror microscope images is rather complicated, only test specimens arc selected having a known composition and topography. Most of the pictures presented in this chapter arc therefore vacuum deposited layers on accurately polished glass disks. A first conductive layer makes the glass surface coincident with an equipotential plane, whereas additional layers, mostly evaporated through a grid of known dimensions, provide a regular pattern. The advantage of using a regular pattern is found in the easy determination of image distortion and magnification. Although in principle discrimination between electrostatic and topographic contrast is possible (conclusion 4 of Section 2.3.3), no clear practical evidence is found yet in the images obtained. Due to the high sensitivity for slight differences in height, the possibility of preparing a specimen with purely electrostatic contrast appears to be rather doubtful. In view of this difficulty it was decided to concentrate primarily most of our efforts on the focused imaging of specimens with topographic contrast. The photographic results presented are only meant to demonstrate the remarkable improvement in image quality of the focused mirror microscope in comparison with the results from mirror projection microscopes. The authors are aware that this series of photographs only provides a limited outlook at the large, but hardly explored, field of possible applications. The achieved improvement in image quality and resolving power of this type of mirror electron microscope,

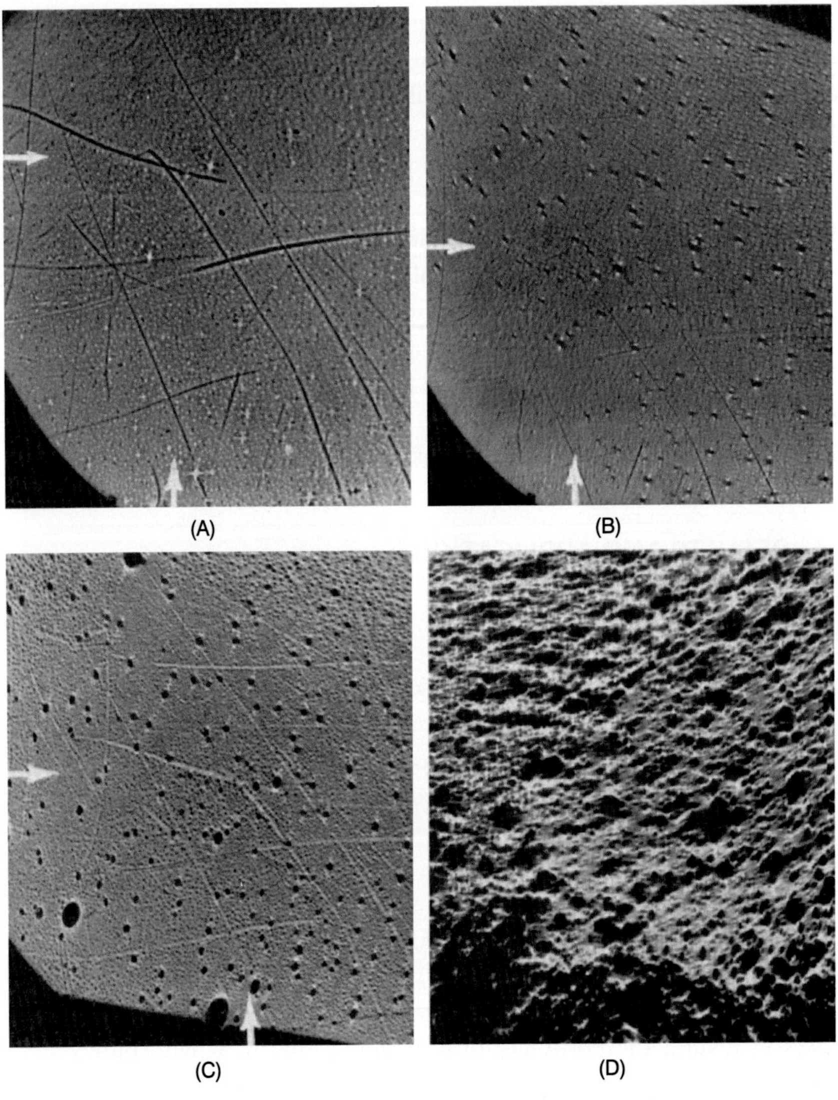

Figure 26 (A) Under-focus (distance off focus −50 μm). (B) In-focus. Magnification 600×. (C) Over-focus (distance off focus +50 μm). (D) Same specimen area after a spark-over to the specimen surface.

the main purpose of this instrument, makes it worthwhile to initiate a more systematic research for widening the scope of useful applications.

Figs. 26A, B, and C represent a through focal series of a vacuum deposited gold layer (30 nm thick) on a polished glass disk. The magnification

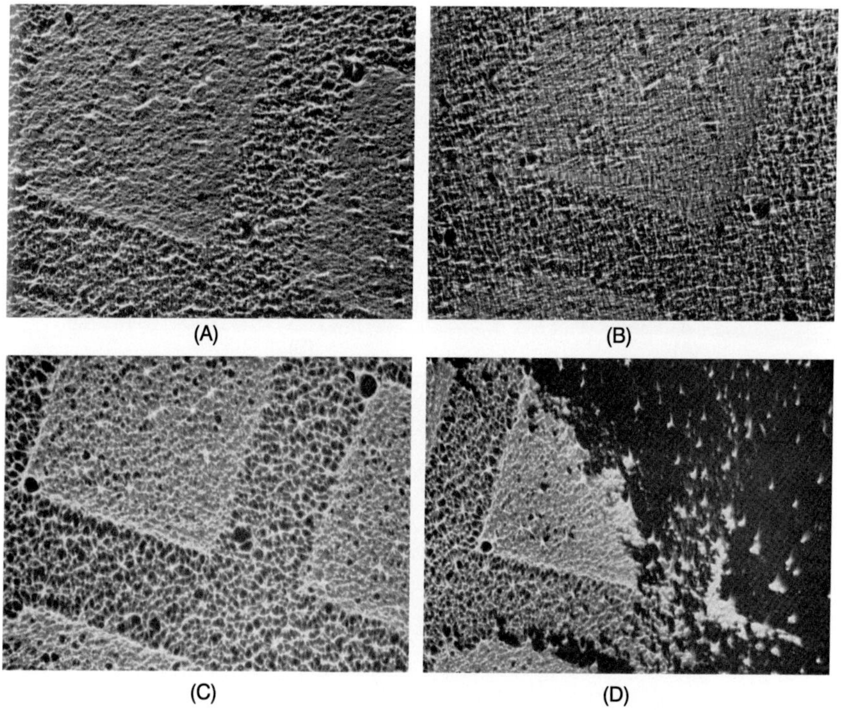

Figure 27 (A) Under-focus (distance off focus −5 μm). (B) In-focus. Magnification 2700×. (C) Over-focus (distance off focus +5 μm). (D) Same condition as Fig. 27C with tilted illumination.

amounts to about 600× and the diameter of the contrast aperture was 200 μm. All white stars in the under focus image clearly change into black bubbles in the over focus image (reverse of contrast). The point of intersection of the arrows indicates the same specimen spot. The occurrence of these local perturbations is mainly due to a continuous ion bombardment of the specimen surface. Uncovered areas of the glass surface, having a poor surface conductivity, cause numerous local charges. Slight differences in the surface conductivity, for example by the bombardment of ions, then lead to a fluctuating potential distribution across the specimen. This phenomenon gives at the final image the impression of the specimen being "alive". The mechanical scratches in the thin gold film fade away for the in-focus condition (Fig. 26B). This effect illustrates the statement, made in Section 2.5, that the perturbed mirror electrode can be considered as a phase specimen.

Figs. 28A, B, and C demonstrate the effect of an increasing tilt of the illuminating beam. Accurate positioning of the contrast aperture in the

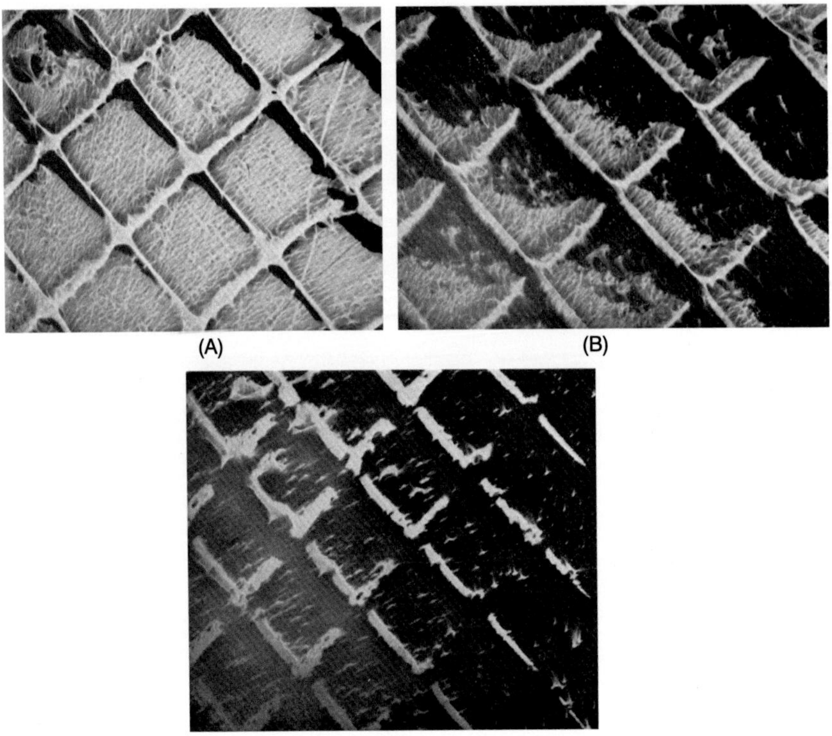

Figure 28 (A) Slightly tilted illumination magnification 900×. (B) Tilted illumination (same specimen area as Fig. 28A).

back focal plane of the combination objective and electrostatic lens gives a knife edge filtering for the entire illuminated area at the specimen. The specimen pictured is the same as in Fig. 27. The magnification is 1000×. The inclined incidence of the illuminating beam causes a narrowing of the bars (actual width about 8 μm) between the squares. The increasing concave mirror action of the bars for tilted illumination provides sharp line foci in the final image.

Fig. 29 is the same specimen as pictured in Fig. 27 but at a magnification of 250×. In spite of the low magnification hardly any image distortion is present. This image is again equally sharp across the entire field of view (about 600 μm).

The in-focus specimen pictured (Fig. 30) is an accurately polished stainless steel surface with copper squares evaporated through a 400 mesh grid (the squares are about 33 × 33 μm^2). The thickness of the copper squares

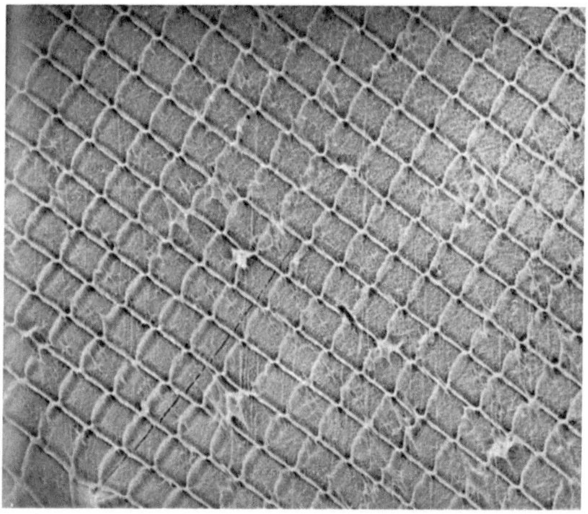

Figure 29 Minimum magnification (250×).

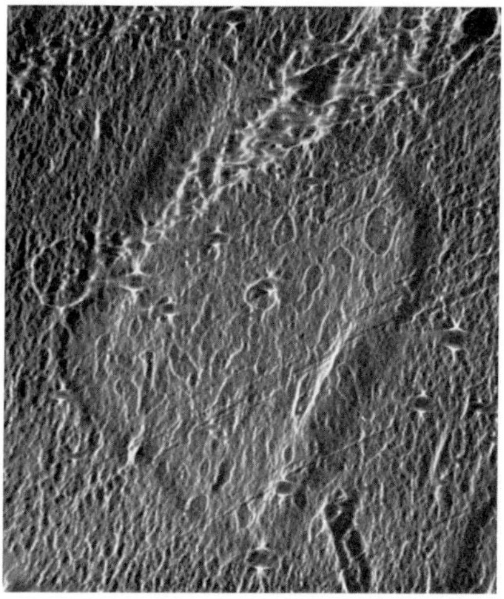

Figure 30 Electrostatic and topographic contrast of a stainless steel surface covered with a copper square (magnification 2500× ± 15%).

amounts to (10 ± 3) nm. The magnification, which is not equal in all directions owing to astigmatism of the deflection bridge, is 3000× ± 15%. The traverse lines are scratches from polishing. The sharply imaged "groove" around the copper square shows a more gradual change in contrast than found for purely topographic specimens (Figs. 27 and 28).

Figs. 31A and B show droplets of aquadag on a gold evaporated glass surface (A shows the in-focus image and B the slightly under-focus con-

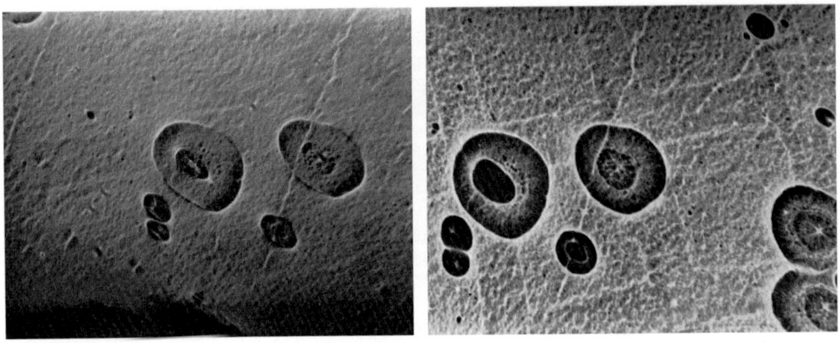

(A) (B)

Figure 31 (A) In-focus (1500×). (B) Under-focus.

dition). The magnification is 1700×. Much detail is revealed within the droplets. A slight amount of astigmatism of the objective lens or caused by the specimen itself is visible in the lower left region of Fig. 31A. It is often difficult from observation of the fluorescent screen to correct entirely for residual astigmatism of the objective lens.

4.2 Applications

The applications mentioned in brief hereafter are not meant to provide the reader with a complete survey about all possibilities of the mirror electron microscope. It only presents a number of applications which might be of interest to physicists investigating surface phenomena at a microscopic scale.

4.2.1 The Investigation of Semi-Conductor Electronics Micro-Circuits

Apart from the surface topography, electric properties as potential distributions across resistors, condensers, etc., and current flow in separate components can be observed. Especially the visualization of the dynamic behavior of micro-circuits allows for determination of interruptions and breakdowns in the circuitry (Igras & Warminski, 1965, 1966, 1967; Maffit & Deeter, 1966; Ivanov & Abalmazova, 1968).

4.2.2 The Investigation of Surface Conductivity, Diffusion of Metals (Igras & Warminski, 1967), and Ferro-Electric Domain Patterns (Igras, Spivak, & Zheludev, 1959)

The movement of electric charges across surfaces, having a poor surface conductivity, can be studied dynamically. Measuring the propagation velocity of electric charges provides information about the surface conductivity

(Mayer, 1960). The storage of charges on photo-sensitive layers (image intensifiers) can be visualized at a high magnification. When the mirror electrode is replaced by a photo-sensitive layer, an image intensifier with a high resolving power could perhaps be realized. Changes in the local work function, resulting from diffusion of metals or doping effects, lead to current density modulations in the final image.

4.2.3 The Investigation of Thin Films

The high sensitivity for topography and local charges offers the possibility to test the quality of evaporated layers. Contaminations and impurities can be easily detected.

4.2.4 The Dynamic Observation of Magnetic Domain Patterns (Spivak et al., 1955; Mayer, 1959a, 1959b)

This is, for instance, the imaging of patterns recorded on magnetic tape (Mayer, 1958; Spivak, Ivanov, Pavluchenko, Sedov, & Shvets, 1963) and magnetic stray fields on grain boundaries (Mayer, 1959a). Some experimenters have successfully reported on observations of ferro-magnetic phenomena (Spivak et al., 1955).

4.2.5 The Investigation of the Local Work Function, as Already Performed in the Emission Electron Microscope

In addition to the mirror images, secondary emission images are obtainable by bombarding the specimen with low energy electrons (in the order of tens of electron volts). This feature provides the possibility to obtain two different types of image from the same specimen area. A stable 100 V source on top of the accelerating voltage is under construction. In how far low energy electron diffraction (LEED) is possible in this instrument remains to be seen.

4.2.6 An Improvement of the Vacuum Near the Mirror Electrode and a Rough Filtering of the Illuminating Beam Yield Interesting Perspectives for Investigation of Physi- and Chemisorption Phenomena

Contrast will be obtained in this case by changes in the local work function resulting from adsorption.

Apart from paying more attention to the practical application of our mirror electron microscope, the following instrumental improvements are planned for the near future:

1. filtering of the illuminating beam by means of a Wien filter,
2. improvement of the vacuum near the specimen.

APPENDIX A. A SCANNING MIRROR ELECTRON MICROSCOPE WITH MAGNETIC QUADRUPOLES (BOK ET AL., 1964)

Le Poole's basic idea of using two crossed quadrupoles for a scanning mirror electron microscope originated as a side line during the process of designing a transmission electron microscope with magnetic quadrupoles (Le Poole, 1964b). From the theory of quadrupoles it is known that a single quadrupole has small coefficients of spherical aberration and, moreover, that it offers more possibilities for correction than rotationally symmetric lenses. However, since the formation of a real image requires at least two quadrupoles arranged in such a way that they counteract each other to a large extent, the actual errors for such a doublet or triplet are of the same magnitude or larger than those of the rotationally symmetric lenses. The new idea was, that when anamorphotic imaging (different magnifications in perpendicular directions) is accepted, a quadrupole doublet can be made which possesses a highly reduced spherical aberration. Since calculations revealed that the formation of a focal line of 0.1 nm in width should be possible, this idea might involve a new method to improve the resolving power for a transmission electron microscope. Obviously good resolution is only required in the direction of the line since no electrons hit the specimen outside the line focus of the illuminating system. Sufficient field of view is obtained by scanning the line focus across the specimen.

As objective lens an identical quadrupole doublet can be used on the same axis but rotated over 90 degrees. If the specimen is replaced by a mirror both doublets are combined in one. The lens action of a magnetic quadrupole changes over 90 degrees for reversed directions of the electron beam.

The irreversible lens action of these lenses thus provides an elegant method for designing a mirror electron microscope for focused images.

In the summer of 1963 the design of a scanning mirror electron microscope was started in collaboration with J. Kramer.

In Fig. 32 the ray diagram for the transmission system is presented. For the scanning mirror instrument, part BQ of the z-axis should be rotated

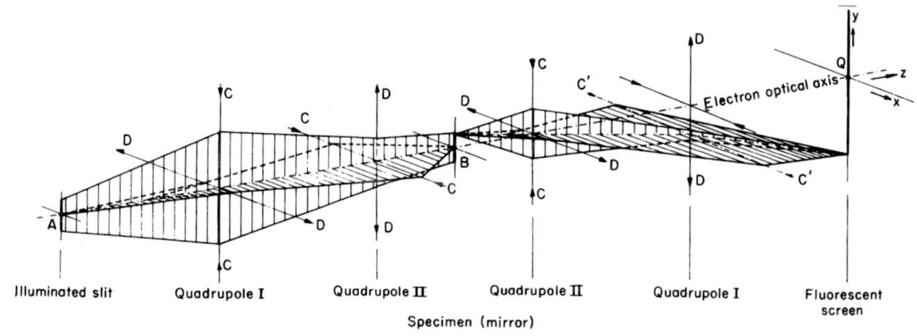

Figure 32 Ray diagram (the specimen is at *B*).

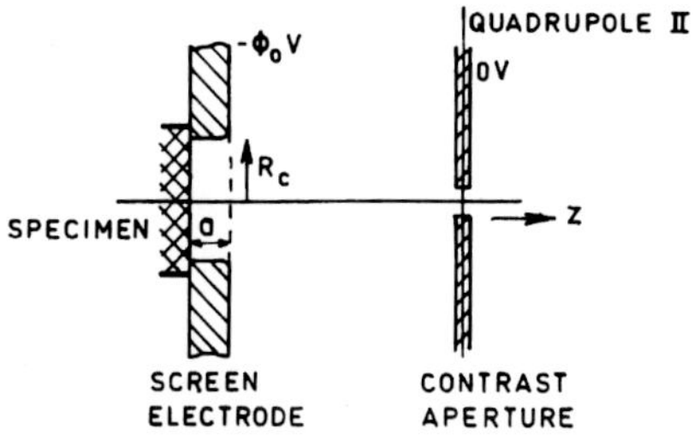

Figure 33 Mirror electrode configuration.

around *B* (the specimen) till *Q* and *A* coincide. The convergent and divergent lens action of the quadrupoles are marked with *C* and *D*.

The anamorphotic imaging on the specimen (mirror) of an illuminated aperture slit at *A* gives a reduction of roughly 100× in width and 15× in length. For the reflected beam this effect reverses, so the length is multiplied by 100× and the width by 15×. In this case too, sufficient field of view is obtained by sweeping the focal line across the specimen by means of two deflection systems. The small amplitude of this movement at the specimen is again magnified at the screen. Proper excitation of the deflection systems provides equal magnification in perpendicular directions at the screen. One double deflector wobbles the illuminating beam, without lateral displacement, in the aperture slit at *A*. A second deflector, mounted around quadrupole I, directs the deflected beam backward through an aperture located in quadrupole II. In order to avoid, for a larger field of view, vignetting of this contrast aperture for reflected electron pencils, the retarding field is shaped to act like a concave mirror. The contrast aperture, which is adjustable along the *z*-axis, is made coincident with the center of

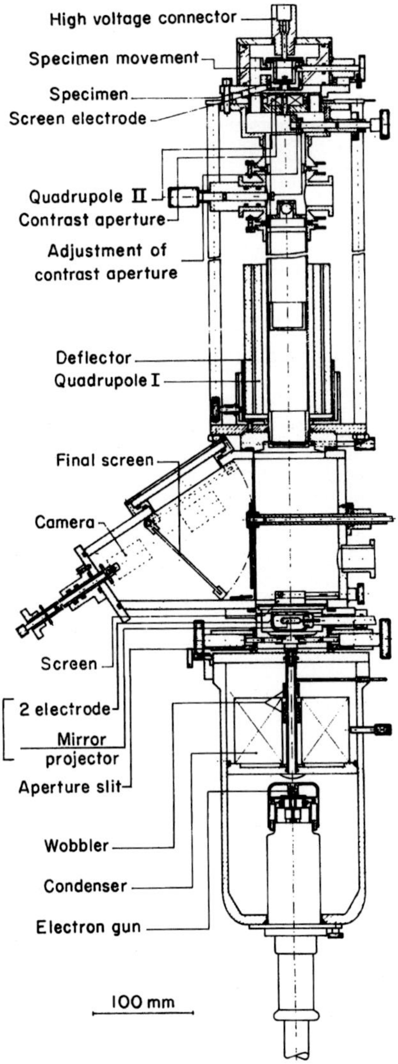

Figure 34 Cross-section through the scanning mirror microscope.

curvature of the concave mirror. The concave mirror action of the retarding field is obtained by using a cup-shaped screen electrode (see Fig. 33). The focal length of this mirror changes for variations of the ratio a/R_c. A slightly off axis two electrode mirror (Le Rütte, 1952), which is mounted below the fluorescent screen projects part of the image with an additional magnification of 20× (total 2000×) onto the obliquely positioned final screen.

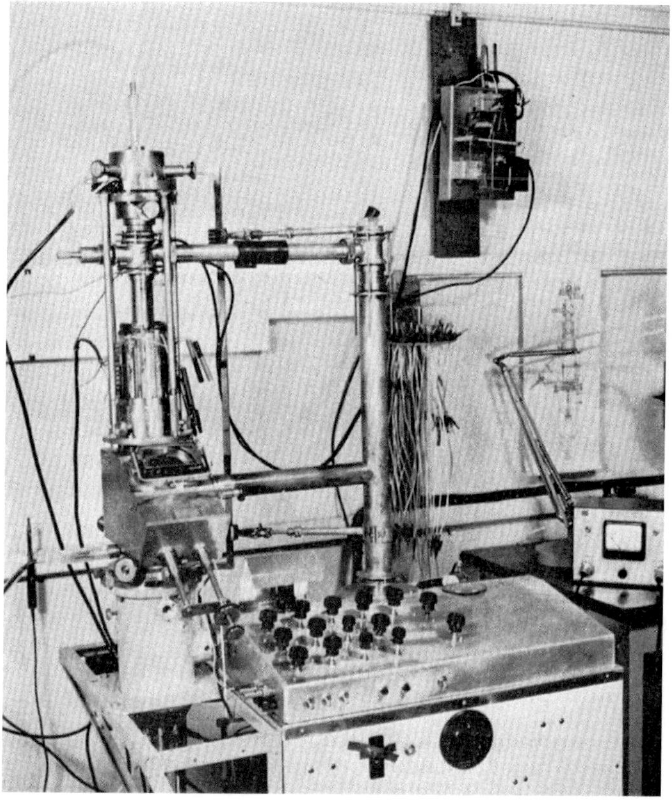

Figure 35 Scanning mirror microscope with quadrupoles.

Fig. 34 gives a cross-section through the instrument, Fig. 35 an overall photograph and Figs. 36A and B the best results obtained.

The specimen pictured is a well polished brass surface covered by ail aluminum layer vacuum deposited through a 750 mesh (33 μm) grid. The magnification, although not perfectly equal in perpendicular directions, is in the order of 1200×. Since a further improvement of the image quality with this instrument could only have been achieved with an entirely redesigned apparatus (better shielding against stray magnetic fields, better mechanical and electrical stability), it was then decided (summer 1964) to leave the idea of magnetic quadrupoles and to change over to a new mirror electron microscope with rotationally symmetric lenses and a magnetic prism.

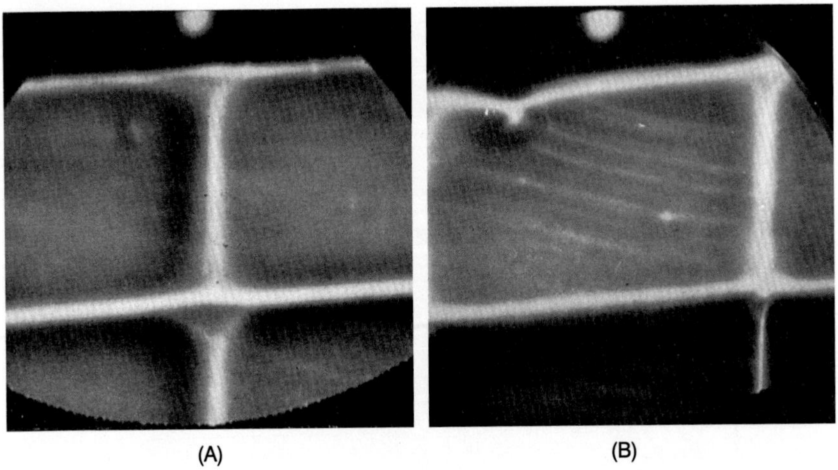

Figure 36 Brass surface with vacuum deposited aluminum squares. (A) Focused on the brass "bars", (B) focused on the specimen surface (topography). Magnification 1200×.

APPENDIX B. CALCULATION OF THE INFLUENCE OF THE SPECIMEN PERTURBATION ON THE PHASE OF THE REFLECTED ELECTRON BEAM IN A MIRROR ELECTRON MICROSCOPE

H. de Lang
Philips Research Laboratories, Waalre, Netherlands

The influence of the specimen perturbation on the phase of the reflected wave will be calculated making use of the concept of phase length along optical rays. Since the axial component of the electron velocity goes through zero near the plane of the mirror (specimen), the wave length of an axial beam becomes infinite there. It might seem doubtful, therefore, if the simple concept of ray-optical phase length would make any sense here. We will show, however, that such an approximation is justified.

B.1 Justification of the Use of the Ray-Optical Phase Length in Relation to Calculations Near the Specimen in a Mirror Electron Microscope

Suppose a plane monochromatic wave directed along the z-axis (normal to the specimen) enters the homogeneous retarding mirror field. The reflected wave front shows a phase ripple with an amplitude equal to the modulation amplitude of the phase length, provided that the interaction area near the specimen is sufficiently "thin". By two different arguments it will be demonstrated that this condition is fulfilled for this specific problem. In the first case use is made of the uncertainty principle, in the second the radius of the first Fresnel zone is compared with the perturbation period.

1. *The Uncertainty Principle.* The following assumptions are involved:
the specimen possesses a sinusoidal perturbation with spatial frequency ν and amplitude Δz;
the axial thickness of the effective interaction area amounts to $(\pi\nu)^{-1}$;
the lateral uncertainty in location for an electron entering the interaction area is a fraction of ν^{-1} (say $(8\nu)^{-1}$).

We are allowed to use the ray-optical phase length if the lateral wandering, as a result of the lateral electron velocity, is also within a fraction of ν^{-1} (say again $(8\nu)^{-1}$).

From the uncertainty principle for the electrons entering the retarding mirror field

$$\Delta y \Delta v_y \approx h/m \tag{47}$$

where

$$\Delta y = (8\nu)^{-1}$$

It follows that

$$\Delta v_y \approx 8\frac{h\nu}{m} \tag{48}$$

Multiplying Δv_y with the time the electron travels through the interaction area, the lateral electron wandering then found must not exceed $(8\nu)^{-1}$ which yields

$$4\left(\frac{z_2}{\pi\nu}\right)^{\frac{1}{2}} v_2^{-1} 8\frac{h\nu}{m} < (8\nu)^{-1}$$

or

$$\nu^{-1} > 26.5\left(z_2\lambda_2^2\right)^{\frac{1}{3}} \tag{49}$$

Taking $z_2 = 3 \times 10^{-3}$ m and $\lambda_2 = 7 \times 10^{-12}$ m (30 kV) we find

$$\nu^{-1} > 140 \text{ nm} \tag{49a}$$

which means that ν does not need to be much smaller than the practical resolving power in the lateral direction.

2. *The Radius of the First Fresnel Zone Compared with the Perturbation Period.* After traveling through the interaction area the radius of the first Fresnel zone has to remain within a fraction of ν^{-1} (say $(8\nu)^{-1}$).

The wave length $\lambda = \lambda(z)$ is

$$\lambda = \frac{h}{m}\left(v_y^2 + v_z^2\right)^{-\frac{1}{2}}$$

Subtracting the unperturbed phase-length:

$$2\pi\, dzh^{-1} mv_z$$

from the perturbed one:

$$2\pi\, dzh^{-1} m\left(v_y^2 + v_z^2\right)^{\frac{1}{2}} \cos^{-1}\alpha = 2\pi\, dzh^{-1} m\left(v_y^2 + v_z^2\right) v_z^{-1}$$

leads to:

$$2\pi\, v_y^2 h^{-1} m v_z^{-1} dz = 2\pi\, v_y^2 h^{-1} m\left(\frac{z_2}{z}\right)^{\frac{1}{2}} v_2^{-1} dz.$$

Integration of this expression over the interaction path must result in π. Thus we obtain

$$8\pi^{\frac{1}{2}} v_y^2 h^{-1} m z_2^{\frac{1}{2}} v_2^{-1} \nu^{-\frac{1}{2}} = \pi \tag{50}$$

The requirement that the radius of the first Fresnel zone (equal to the lateral velocity times electron traveling time) be smaller than $(8\nu)^{-1}$ gives

$$v_y < \pi^{\frac{1}{2}} \nu^{-\frac{1}{2}} v_2 z_2^{-\frac{1}{2}}/32 \tag{51}$$

Combining Eqs. (50) and (51) we obtain

$$\nu^{-1} > 17.3\left(z_2 \lambda_2^2\right)^{\frac{1}{3}} \tag{52}$$

Thus we may conclude that consideration of the radius of the first Fresnel zone (Eq. (52)) as well as the application of the uncertainty principle (Eq. (49)) both justify the use of a simple ray-optical phase-length treatment for ν-values occurring in practice. This greatly facilitates the calculation of the phase modulation in the reflected wave.

B.2 Calculation of the Phase Modulation in the Reflected Wave as a Function of the Amplitude Δz and the Spatial Frequency ν of a Sinusoidal Specimen Perturbation

The unperturbed wavelength as a function of z is

$$\lambda(z) = \left(\frac{z_2}{z}\right)^{\frac{1}{2}} \lambda_2$$

where $(\frac{z_2}{z})^{\frac{1}{2}}$ is the "refractive index".

In order to take into account the perturbation by the specimen we select two rays, i.e. one to a "peak" of the specimen and another to a "valley".

Along the rays to the "peaks" the perturbed wavelength λ' is

$$\lambda'(z) = \lambda\left(z - \Delta z\, e^{-2\pi \nu z}\right) = \lambda_2 z_2^{\frac{1}{2}} \left(z - \Delta z\, e^{-2\pi \nu z}\right)^{-\frac{1}{2}} \tag{53}$$

Along the rays to the "valleys" the perturbed wavelength λ'' is

$$\lambda''(z) = \lambda\left(z - \Delta z\, e^{-2\pi \nu z}\right) = \lambda_2 z_2^{\frac{1}{2}} \left(z + \Delta z\, e^{-2\pi \nu z}\right)^{-\frac{1}{2}} \tag{54}$$

The amplitude R_ζ of the phase modulation is the reflected wave at the plane $z = \zeta$ is

$$R_\zeta = 2\pi \left\{ \int_{-\Delta z}^{\zeta} \left(\lambda''\right)^{-1} dz - \int_{\Delta z}^{\zeta} \left(\lambda'\right)^{-1} dz \right\} \tag{55}$$

With Eqs. (53) and (54), Eq. (55) transforms into

$$R_\zeta = 2\pi \lambda_2^{-1} z_2^{-\frac{1}{2}} \left\{ \int_{-\Delta z}^{\zeta} \left(z + \Delta z\, e^{-2\pi \nu z}\right)^{\frac{1}{2}} dz - \int_{\Delta z}^{\zeta} \left(z - \Delta z\, e^{-2\pi \nu z}\right)^{\frac{1}{2}} dz \right\} \tag{56}$$

In order to calculate the final value R of the amplitude of the phase modulation we have to take $\zeta = \infty$ in (56). Assuming further that $\Delta z \ll (\nu)^{-1}$ we can easily evaluate the integrals in (56). In doing so we obtain for the final amplitude R of the phase modulation in the reflected wave

$$R = 2^{\frac{1}{2}} \pi \lambda_2^{-1} z_2^{-\frac{1}{2}} \nu^{-\frac{1}{2}} \Delta z \tag{57}$$

Remarks. For a distance $z = (\pi \nu)^{-1}$ from the specimen the reflected wave has already received 95% of its total modulation. Thus the supposed thickness $(\pi \nu)^{-1}$ for the interaction area is in fact reasonable. A similar calculation holds for an electrostatically perturbed specimen. In that case Δz in Eq. (57) has to be replaced by $\Delta \phi / F_z$.

B.3 Visibility of Small Differences in Specimen Relief (Axial Resolving Power)

For a weakly modulated specimen (sinusoidal perturbation) the reflected wave consists of a direct beam (zero order) and both diffracted beams of the first order. The relative amplitude of the diffracted beam with respect to the direct beam amounts to (see Eq. (57)):

$$\frac{1}{2}R = 2^{-\frac{1}{2}}\pi\lambda_2^{-1}z_2^{-\frac{1}{2}}\nu^{-\frac{1}{2}}\Delta z \tag{58}$$

If with a knife edge aperture one of the first order beams is intercepted, the remaining first order beam and the direct beam give in the imaging plane an interference pattern with a sinusoidal intensity modulation. The modulation depth is

$$(I_{\max} - I_{\min})/(I_{\max} + I_{\min}) = R = 2^{\frac{1}{2}}\pi\lambda_2^{-1}z_2^{-\frac{1}{2}}\nu^{-\frac{1}{2}}\Delta z \tag{59}$$

Taking a modulation depth of 0.1 as the threshold value, the corresponding minimum specimen ripple amplitude then amounts to

$$(\Delta z)_{\min} = 2.2 \times 10^{-2}\lambda_2 z_2^{\frac{1}{2}}\nu^{\frac{1}{2}} \tag{60}$$

Taking

$$\begin{aligned} \lambda_2 &= 7 \times 10^{-12}\ \text{m (30 kV)} \\ z_2 &= 3 \times 10^{-3}\ \text{m} \\ \nu^{-1} &= 3 \times 10^{-7}\ \text{m} \end{aligned}$$

the threshold value of specimen relief $(\Delta z)_{\min}$ is found to be 16×10^{-12} m.

B.4 Influence of the Energy Spread of the Incident Electron Beam on the Axial Resolving Power

The results obtained in the previous sections are only valid for a monoenergetic electron beam. To approximate the influence of an energy spread a rectangular energy distribution (width $\Delta\phi$) is supposed to be present in the incident beam.

The contribution to the image intensity of a narrow interval $d\phi$ out of the energy width $\Delta\phi$ is represented by

$$dI = d\phi\left(A + Be^{-2\pi(d\phi/\phi_2)\nu z_2}\sin 2\pi\nu y\right), \tag{61}$$

where A, B are constants.

Integrating Eq. (61) from zero up to $\Delta\phi$ we find

$$I = A\Delta\phi + B\Delta\phi \frac{\phi_2}{2\pi\nu z_2\Delta\phi}\left(1 - e^{-2\pi(\Delta\phi/\phi_2)\nu z_2}\right)\sin 2\pi\nu y$$

For the factor f by which the modulation depth of the image intensity is reduced as a result of a rectangular energy spread $\Delta\phi$ we find

$$f = \frac{\phi_2}{2\pi\nu z_2\Delta\phi}\left(1 - e^{-2\pi(\Delta\phi/\phi_2)\nu z_2}\right) \tag{62}$$

Taking

$$\left.\begin{aligned} \phi_2 &= 3\times 10^4\ \text{V}\\ z_2 &= 3\times 10^{-3}\ \text{m}\\ \nu^{-1} &= 3\times 10^{-7}\ \text{m}\\ \Delta\phi &= 0.5\ \text{V}\end{aligned}\right\}\quad \text{we find } f = 0.55$$

Conclusion. The minimum detectable specimen relief as expressed by (60) and (62) is three orders of magnitude smaller than the "ad hoc" value $\frac{\Delta\phi}{F_z}$. In fact the influence of the energy spread $\Delta\phi$ on the axial resolving power is not severe as long as the energy spread does not exceed the thickness (in terms of energy) of the interaction layer. It might seem paradoxical that such a favorable axial resolving power is possible with non-monochromatic radiation. It must be pointed out, however, that with interference microscopes using broadband visible light, observation of depth differences several orders of magnitude smaller than the wave length is quite common.

APPENDIX C. PRACTICE OF MIRROR ELECTRON MICROSCOPY

H. Bethge, J. Heydenreich
Institut für Festkörperphysik und Elektronenmikroskopie, Halle, Germany

C.1

Since the first investigations in mirror electron microscopy almost 35 years ago, the philosophy of design of the devices and the application of this method have changed considerably. The first mirror electron microscopes (Hottenroth, 1937; Orthuber, 1948; Bartz & Weissenberg, 1957; Bartz, Weissenberg, & Wiskott, 1952) were devices with a simple magnetic prism, some of which were glass devices. In order to avoid the difficulties connected with the astigmatism of the magnetic deflection field, most of the later mirror electron microscopes were straightforward devices without a magnetic prism, that means with the same beam axis for the incoming as well as the reflected electrons (see e.g. Mayer, 1955; Bethge et al., 1960; Spivak et al., 1961).

Devices of this type have proved very good and are also nowadays successful in operation. Nevertheless, they have some disadvantages, the most essential of which are: 1. The center of the image cannot be observed, because the aperture for the passage of the primary beam is in the way. 2. The manipulation of the beam of reflected electrons, e.g. a further magnification by a projector lens, is not possible without influencing of the primary electron beam. 3. With the aid of such a straightforward device, only shadow images can be obtained and not focused images (see e.g. Schwartze, 1965). Since possibilities exist for the production of deflection fields without noticeable astigmatism (see e.g. Archard & Mulvey, 1958), an increasing number of mirror electron microscopes are being produced with a magnetic prism (see e.g. Mayer et al., 1962; Schwartze, 1966; Bok, 1968).

C.2

An example of a straightforward electron optical device with the same beam axis for the primary electrons and the reflected electrons, is given in Fig. 37, which is a sectional view of a mirror electron microscope. This is described more in detail by Bethge et al. (1960). Viewing from bottom to top, the microscope column consists of the following main parts: electron source, viewing chamber, projection tube, and specimen chamber.

The electron gun is a simple triode system (T), and possibilities for mechanical alignment (A) of the electron beam are included. The primary beam, after having passed two apertures in the viewing chamber, reaches the specimen (O), where it is reflected. The actual electron mirror system consists of the specimen (O), mounted on an insulated table and aperture (H) placed in front of it. This system acts as a diverging mirror. The resulting "magnification" on the screen (S), is affected firstly by the size of the aperture (H) and secondly by the distance between the specimen and the aperture. The specimen holder can be shifted horizontally under vacuum by the aid of the screw S_1; the distance specimen–aperture is also changeable under vacuum (screw S_2). A further possibility for changing the magnification is given by the lens action of two cylindrical electrodes (C). The screen

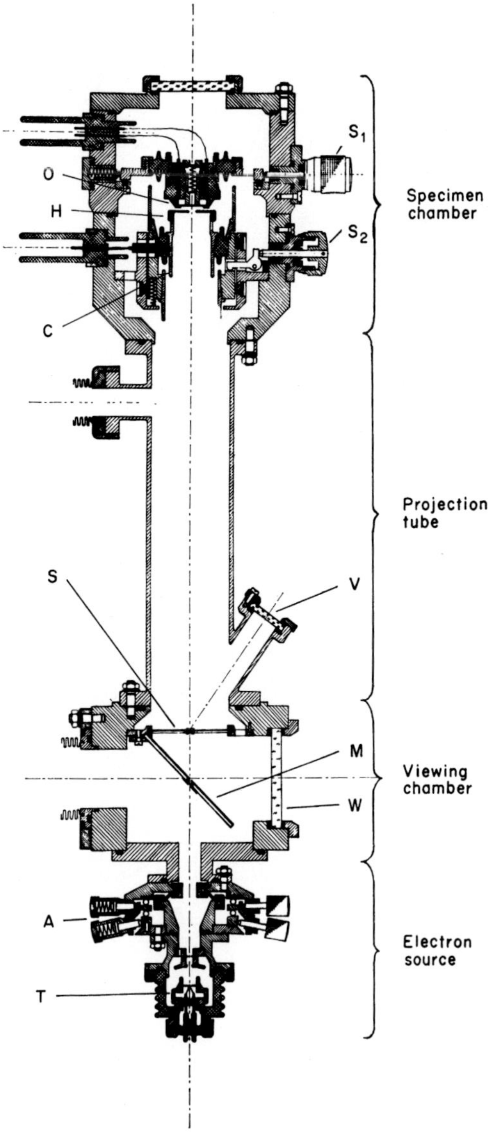

Figure 37 Sectional view of a mirror electron microscope without beam deflection.

(S) can be observed either through the small viewing hole (V) at the side or through a window (W) by the aid of a mirror (M) inclined at 45 degrees. Photographs are taken from outside by the aid of a standard camera placed in front of this window. For changing the specimen, the upper part of the

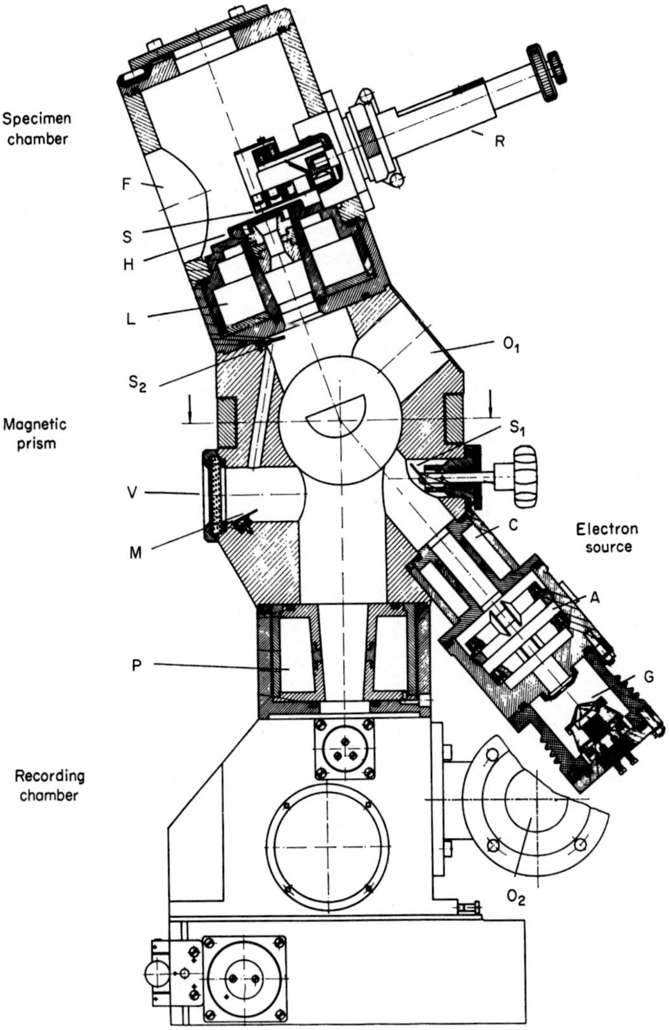

Figure 38 Sectional view of a mirror electron microscope with a magnetic prism.

specimen chamber is removed. The beam voltage can be varied between 4 and 15 kV; the resulting maximum magnification is about 300×.

C.3

As an example of a modern mirror electron microscope with a magnetic prism, a device described in detail by Heydenreich (1970) is shown in a sectional view in Fig. 38. In the device a deflection angle of 20 degrees is used. The main parts of the microscope (magnetic prism, electron source and specimen chamber) are mounted on top of a recording chamber (containing screen and plates) as used in commercial electron microscopes.[3] The electron source is fitted at the side of the magnetic prism column. Starting from the electron gun (G) the electron beam is aligned with the aid of combined electric and magnetic adjustment equipment (A) and is eventually focused by a condensor lens (C). Adjustment aids are a removable screen (S_1) and a further auxiliary screen (S_2). Both the screens are observed through a special viewing window (V), the screen (S_2) via a small inclined mirror (M). The coils of the magnetic prism are above and below the plane of the drawing; in the sectional view the form of the pole pieces can be seen. The specimen (S), which is insulated with respect to the microscope column, is mounted on a rod (R), which allows the necessary horizontal movement of the specimen during the electron optical investigation to take place. Specimen change is effected through the front opening (F) of the specimen chamber, which is closed with a sealing plate during routine operation. This sealing plate can be replaced by a special attachment for specimen treatments (also not drawn in the sectional view). After having brought the specimen into place in this attachment by shifting the specimen rod, the specimen to be investigated can be exposed to an ion beam for cleaning the surface. Furthermore, this attachment contains evaporation sources for the formation of thin layers to be investigated in the mirror electron microscope. The specimen chamber has been so designed that it will readily accept equipment for cooling, heating, magnetizing or straining the specimen. The aperture (H), necessary for producing a shadow image, is mounted on the objective lens (L), which is important for producing focused images. With the aid of the projector lens (P), shown in the sectional view without pole pieces, a considerable increase in the end magnification of the image is given. The microscope column is pumped out to a vacuum of about 10^{-5} Torr through two openings (O_1, O_2).

The stability of the beam voltage and the currents to be used (including that of the magnetic prism) is about 10^{-5}. The beam voltage can be varied over a wide range; it proved favorable to use voltages between 10 and 20 kV. For the shadow image technique, which is used in the majority of cases, the resolving power is about 1000 Å; the maximum attainable magnification lies in the region of $3000\times$.

[3] In this case: Electron Optical Plant EF (VEB Carl Zeiss Jena).

As is well known the mirror electron microscope is suitable for imaging the relief as well as electric or magnetic inhomogeneities of a specimen surface.

C.4

The imaging of *geometric inhomogeneities* of a specimen surface, of which impressive examples in the early days of this technique have been given by Bartz, Weissenberg, and Wiskott (1954) on metal surfaces, suffers from the considerably restricted resolving power of the mirror electron microscope used in shadow image technique (e.g. Schwartze, 1965, 1965–1966, 1966) which for practical reasons is of the order of 1000 Å. Nevertheless, because of the special type of image formation in the mirror electron microscope, the imaging of the surface relief is of some interest. Since the information about the specimen surface in the imaging electron beam is given by the potential field in front of the object, which is especially sensitive to different "heights" of surface inhomogeneities, a kind of "spatial" detection of the specimen surface is carried out. In this way one gets pronounced shifting effects of image structures, which are larger, the rougher a surface irregularity is. In many cases this effect is regarded as a disturbance of the image and leads to the requirement for smooth specimen surfaces in mirror electron microscopy. On the other hand, this shifting effect of the image structure gives additional information about the specimen surface, in so far as the shifting effect is detectable by comparing different pictures of the same specimen region.

An example of this is given in Fig. 39, which shows a series of pictures of the same region of a NaCl cleavage face (lightly coated with evaporated Ag). In (a) is shown the image taken with the mirror electron microscope, in (b) the corresponding conventional microscope image, and in (c) the electron mirror image of the "matched" cleavage face. Assuming the conventional microscope image (b) is a "true" image of the object, we see cleavage steps partly joining, thereby forming tiplike regions (B, C). Because of the special type of contrast formation in the mirror electron microscope (shadow image technique), each step is imaged as a bright–dark double line, the bright edge of which marks the lower plateau adjoining the step. For this reason the electron optical image (a) shows that regions A and C are depressions and region B is a protrusion. As briefly mentioned in the introduction (see also Heydenreich, 1966) a further consequence of the mechanism of contrast formation is that, especially at rough specimen surfaces, the double lines marking the steps are not found at the position

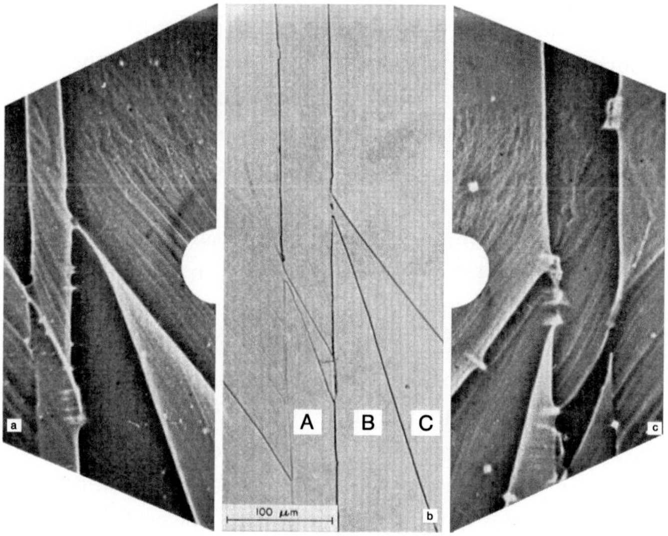

Figure 39 Rough cleave steps on a NaCl cleavage face. (a) and (c) Electron optical micrographs (cleavage face and opposite cleavage face), (b) conventional microscope image.

corresponding to the object according to a conventional microscopic imaging process, but the double lines are shifted in the "image plane" in the direction of the lower plateau. In the example, this effect is easily recognizable from the fact that the lower regions A and C seem to be much smaller in the electron optical image than in the conventional microscope image, and that the plateau region B seems to be correspondingly larger. As can be seen from Fig. 39C the effect is reversed in the "matched" cleavage face. According to interferometric measurements, the height of the steps amounts to about 700 Å. Conclusions about the roughness of the specimen surface are possible from the shift of the image structures, pre-supposing that adequate calibration has taken place, in which careful attention must be paid to specimen potential.

C.5

The pictorial representation of *electrical surface inhomogeneities* with the aid of the mirror electron microscope is of special importance. Naturally, the scanning electron microscope, which is now well established as a commercial device, can also be used for the imaging of electrical specimen inhomogeneities. However, in this case one has to take into account the fact that in the formation of the image a great number of electrons is shot into the specimen surface. This is especially critical in the investigation of semiconductor surfaces, which are the specimens most frequently used for such investigations. In this case the mirror electron microscope, in which there is no direct interaction between the imaging electron beam and the specimen surface, proves a very useful instrument because of the absence of possible specimen modification. Investigations of electrical surface inhomogeneities by the aid of the mirror electron microscope have already been carried out by Orthuber (1948), who was interested in potential distributions in metallic and semi-conducting surfaces. As special examples of the imaging of electric structures can be cited the imaging of p–n junctions, carried out first in 1957 by Bartz and Weissenberg, and the observation of ferro-electric domains (barium titanate) by Spivak et al. (1959a). Of special interest is the pictorial representation of electric conductivity distributions which is possible in the mirror electron microscope by producing a corresponding current flow in the specimen surface (e.g. Mayer, 1957a). As an example of this technique, Fig. 40 shows the electron optical image of a thick germanium layer (900 Å) which is covered by a very small quantity of evaporated indium with an average thickness of about 40 Å. Without any current flow (A) the only image contrasts which are seen are those which are related to the rough surface of the evaporated layer. When a voltage is applied between the left and the right-hand sides of the specimen surface a current flow results and one sees contrast lines of increasing clearness in the images (B), (C), and (D) (respectively: +2, +10, +20 V) which show that strong locally acting electric fields are present. These strong fields are related to the boundaries of the flat indium islands, which are more or less insulated from each other. When current flows through the specimen the more negatively charged areas appear darker in the image of the border region than the more positive areas. In this example the image contrast corresponding to electrical inhomogeneities at large current flows is much stronger than that corresponding to surface roughness which is simultaneously present. By a suitable adjustment of the specimen potential one always succeeds in suppressing the weaker contrast (in the present case, geometrical). Fig. 41 shows that, for the specimen under discussion, the specimen potential necessary for this suppression is between −30 and −40 V.

C.6

In the mirror electron microscopy of *magnetic inhomogeneities*, the first investigations of which were carried out by Mayer (1957b) and by Spivak el al. (1955), the action of the normal component of the magnetic field at the

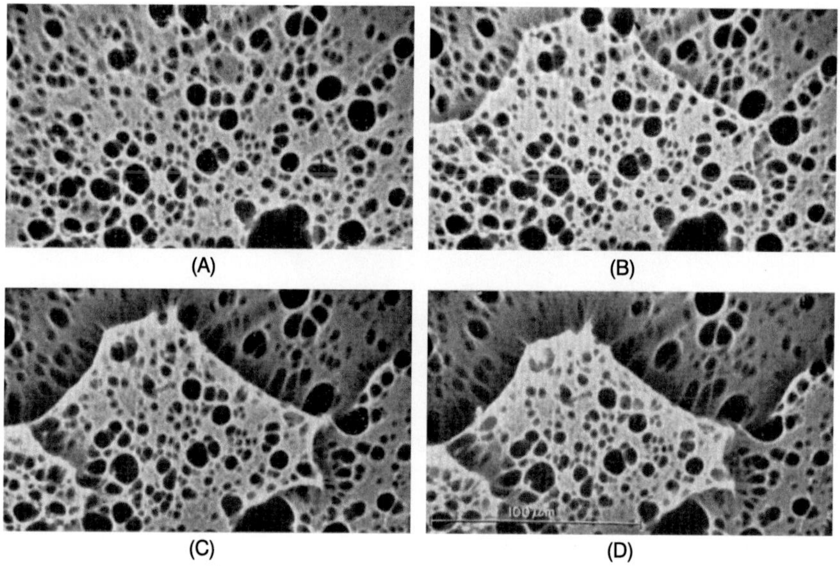

Figure 40 Rough germanium layer, coated with a thin indium layer. Electron optical micrographs with different currents flowing in the layer, produced by different applied voltages: (A) 0 V, (B) +2 V, (C) +10 V, (D) +20 V.

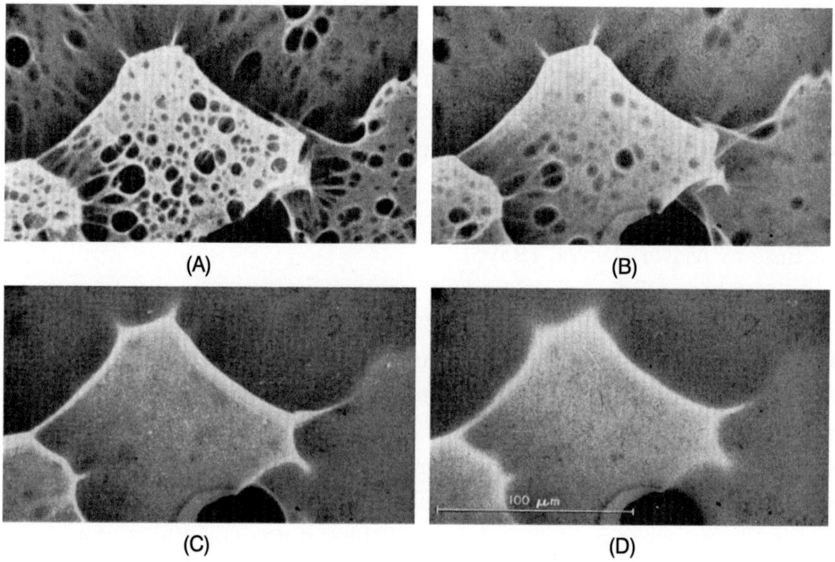

Figure 41 See Fig. 40. Applied voltage for current flow in the layer: +50 V. Specimen potential: (A) −10 V, (B) −20 V, (C) −30 V, (D) −40 V.

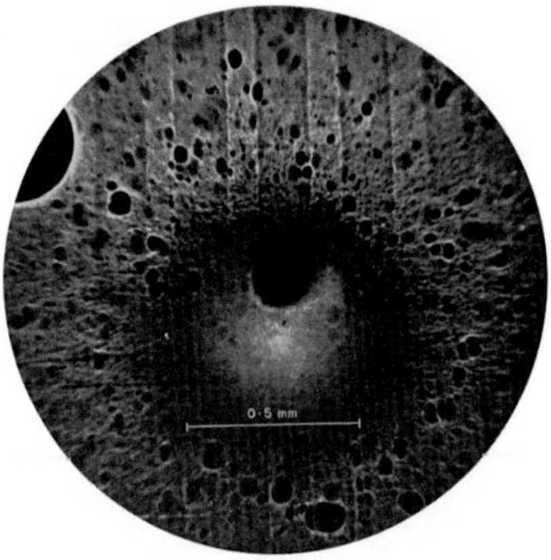

Figure 42 Demonstration of the imaging of magnetic structures: magnetic tape with recordings on it (rectangular pulses).

specimen surface on the radial component of the electron velocity in the reflection plane, is responsible for the formation of contrast. Since, in the usual shadow image technique, the radial components of the electron velocity in the region of the intersection of the beam axis with the specimen surface are zero (or very small) and only increase with increasing distance from the axis, one can only image contrasts from magnetic surface irregularities at a distance from the center of the image. The possibility thus exists of distinguishing between image contrasts which are related to geometric or electric surface inhomogeneities and those which are related to magnetic in homogeneities. The main field of application of the mirror electron microscope for studying magnetic structures is the pictorial representation of magnetic domains, which was successfully achieved by this technique for the first time by Mayer (1959a, 1959b).

A demonstration of the special types of contrast formation in the imaging magnetic structures is given in Fig. 42 which shows a special magnetic tape with recordings (rectangular pulses). At the center of the image (which, in this case, does not lie on the screen center and which is marked by a diffuse spot of electron impact) the magnetic contrast is less pronounced than toward the edge of the screen. Contrast reversal occurs between the upper and lower regions of the specimen resulting from a reversal in the radial component. To the right and left of the image there is no magnetic contrast since, because of the direction of the radial components of the electron velocity, the magnetic force acts in the same direction as the stripe structures in the image. The difference between the widths of the stripes ("recordings") on the left- and right-hand sides of the image, shows that there is also a shifting effect of the magnetic contrast lines which is more pronounced the stronger the magnetic fields. These shifts are in opposite directions on opposite sides of the center of the image and give the impression of different widths of the recordings (rectangular pulses). The true width of the recordings can be obtained by calculating the average.

C.7

Based on the three main possibilities of the mirror electron microscope (detection of geometric, electric and magnetic surface irregularities), the device is being used more and more for special problems of solid state physics. Some examples of these applications will now be briefly reviewed. Attempts to image dislocations in germanium emerging at the surface by the detection of the space charge region around the dislocations were carried out by Igras (1962). With certain restrictions, the local detection of electric microfields in semiconducting surfaces can, as Igras has also shown, be used for example in the investigation of the segregation of impurity atoms, or for the detection of surface drifts. Using a suitable helium cooling stage in the specimen chamber Bostanjoglo and Siegel (1967) were able to image the local distribution of the magnetic flux of superconductors in the intermediate state. Mainly nobium specimens were used; these were kept at a temperature of 3.5 K during the electron mirror observation. According to a brief report by Wang, Challis, and Little (1966), the Abrikosov-vortex structure of some superconducting materials of the type II have also been studied in the mirror electron microscope. In the future solid state circuits and electronic devices in thin film techniques will probably also be studied by mirror electron microscopy.

C.8

Summarizing, it can be said that, in spite of its restricted resolving power, the high sensitivity of the mirror electron microscope in the shadow image technique for electric or magnetic fields on the specimen surface (remembering that the geometry of the surface also influences this field distribution) gives a great deal of information about these properties of the specimen surface. The shifting effect of image structures discussed above gives, on the one hand, additional information about the strength of fields present, as far as it can be detected exactly, but on the other hand this effect causes an uncertainty in the localization of the image structure of these fields. In mirror electron microscopy with focused images, the resolving power is much better than in the shadow image technique and the shifting effects mentioned above are not present, so that there is no problem in image localization. However, this method does not have the high sensitivity to field inhomogeneities of the shadow image method. Magnetic structures, for instance, cannot be imaged by this technique because there are no radial components of the electron velocity. The possibility presents itself of using both techniques in combination so that in the same region of the specimen the localization of the inhomogeneities can be precisely detected in the focused image and the surface fields present are thereafter detected with sufficient differentiation using the shadow image.

In the field of design and development of apparatus, the tendency is toward reliable instruments equipped with a magnetic prism. Special developments such as, for example, the stroboscopic mirror electron microscope (Lukjanow & Spivak, 1966) or the scanning mirror microscope (Garrood & Nixon, 1968) are very useful. For the future it is desirable to develop devices with universal possibilities for specimen treatment. In order to take full advantage of the high sensitivity of the mirror electron microscope for surface potential distributions, it is necessary to aspire to the investigation of clean surfaces (free of adsorption layers). For this reason ultrahigh vacuum devices are becoming of increasing importance in the field of mirror electron microscopy.

APPENDIX D. SHADOW PROJECTION MIRROR ELECTRON MICROSCOPY IN "STRAIGHT" SYSTEMS

M.E. Barnett
Imperial College, London, United Kingdom

Work on mirror microscopy in England was begun in 1962–63 at the Engineering Department of Cambridge University. More recently, the subject has been studied in London University, at University College and Imperial College. Work has been largely confined to the shadow projection mode, using "straight" systems with magnetic lenses. Fig. 43 shows the basic design of system (Szentesi & Barnett, 1969), and is in fact a schematic scale diagram of the instrument in use at present at the Electrical Engineering Department of University College. This is a small, low magnification system, similar in layout to the column originally used in

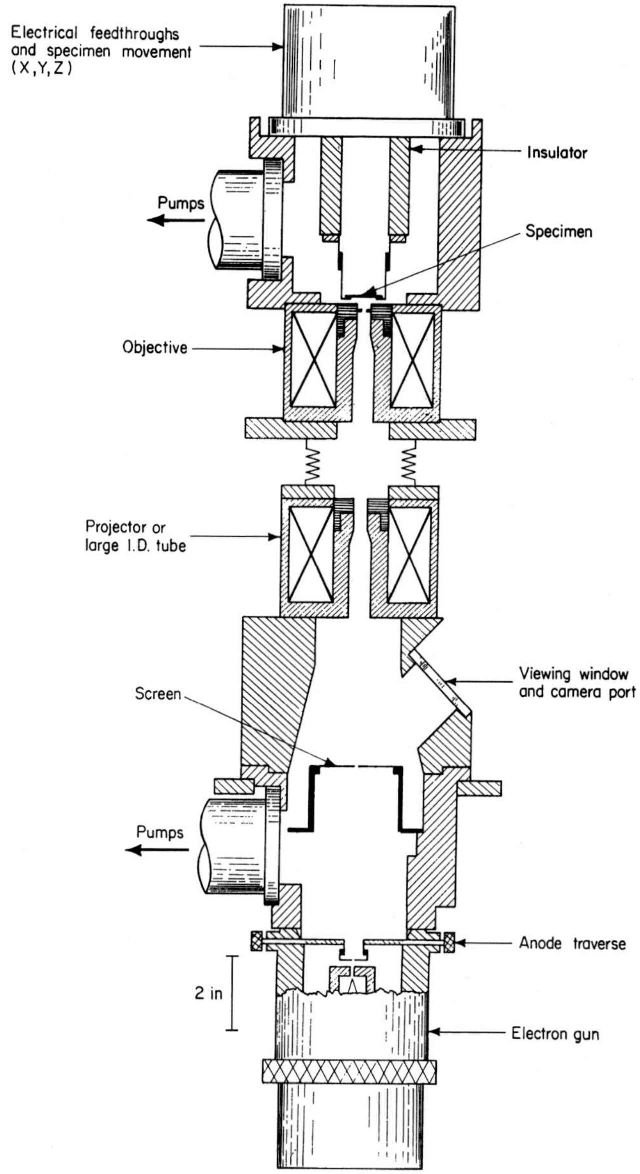

Figure 43 A simple "straight" design of a mirror electron microscope.

Cambridge in 1963. Other columns have been variations on this basic design. A condenser Ions can be placed between gun and screen (Barnett & Nixon, 1964) to improve the image brightness at magnifications greater

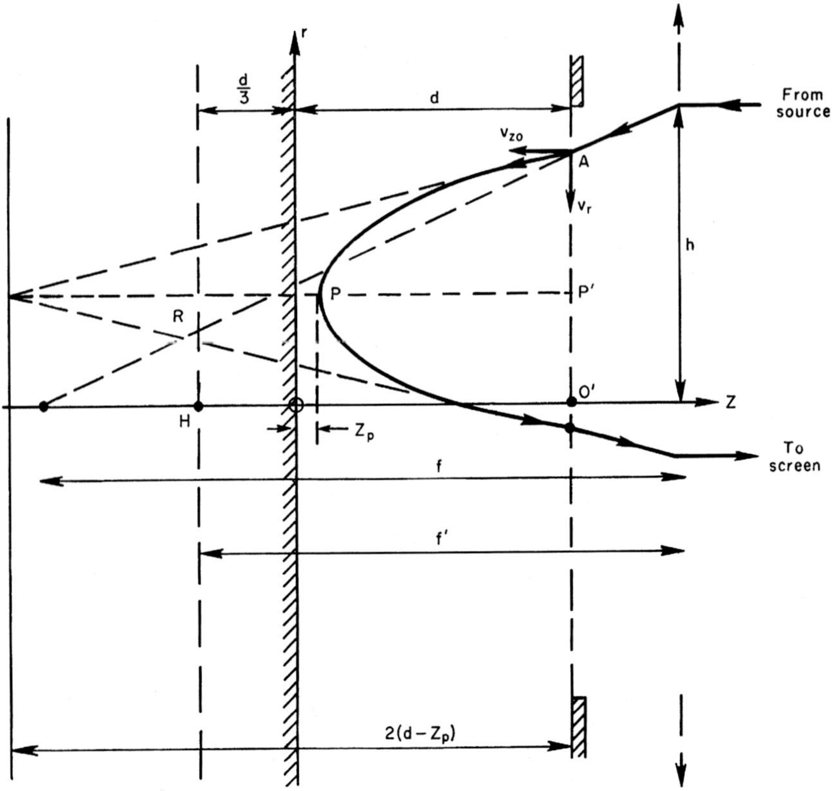

Figure 44 Geometrical ray paths in the objective stage.

than a few hundred. This is necessary since a short focal length projector lens greatly reduces the intensity of the upward beam. In a more elaborate system (Barnett & Nixon, 1967a), designed for magnifications up to 2500×, a two lens projector system, as well as a condenser system was incorporated. In this column, a 45° mirror was located under the transparent viewing screen, so that photography normal to the screen was possible (cf. the micrographs of Fig. 45). In simpler systems the screen is photographed at an oblique angle, hence the foreshortening in most micrographs.

D.1 The Objective Stage

Since the image contrast results from the modification, by the surface microfields, of the radial velocities of the electrons, it is useful to know what the radial velocities at the specimen would be in the absence of microfields, i.e. if the specimen were a nonmagnetic plane equipotential surface. Fig. 44 shows geometrical ray paths in the objective stage assuming a plane specimen (at $z = 0$) and a uniform retarding field between the top surface of the objective lens (at $z = d$) and the specimen. It is easily shown (Barnett & Nixon, 1967b) that the radial velocity of the electrons at their point of reversal is given by

$$v_r = \frac{r}{6d}\left(\frac{2eV}{m}\right)^{\frac{1}{2}}\left\{1 - \frac{8d}{3(f - f')}\right\} \tag{63}$$

where V is the original beam energy, f is the focal length of the objective lens and f' is the distance between the principal plane of the objective lens and the principal plane HR ($z = -d/3$) of the mirror. This expression shows that v_r changes sign at $f = f' + \frac{8d}{3}$ and again at $f = f'$ (the latter value corresponding to the formation of a focused probe at the specimen surface). Reversals of image contrast occur at these values of f.

The existence of radial velocity components in the beam can be shown to cause the beam to be reflected at a paraboloid of revolution rather than along a plane. If the specimen is at a small bias voltage V_b negative with respect to the cathode the locus of reversal points is

$$z = \frac{V_b}{E} + \frac{r^2}{36d}\left\{1 - \frac{8d}{3(f - f')}\right\}^2 \tag{64}$$

E being the field strength at the surface.

The curvature of the reversal surface gives rise to a falling off of sensitivity with increasing distance from the axis. This effect is clearly evident in the series of micrographs in Figs. 45A–D. The specimen is a cleaved rocksalt crystal rendered conducting by the evaporation of a thin layer of gold. The series shows the effect on image contrast of changing the specimen bias voltage; the more positive the specimen, the finer the topographical surface detail revealed. In the final micrograph, the specimen is positive with respect to the cathode, and the central region of the image consists of reflected electrons due to beam impact.

D.2 Image Contrast

An attractive feature of mirror microscopy is its ability to detect electrical potential variations directly. There is, however, the problem of distinguishing between the electrical and topographical origins of image contrast. This can be investigated using specimens of known form (Szentesi & Barnett, 1969). Fig. 46A shows a portion of an 800 Å thick copper electrode pat-

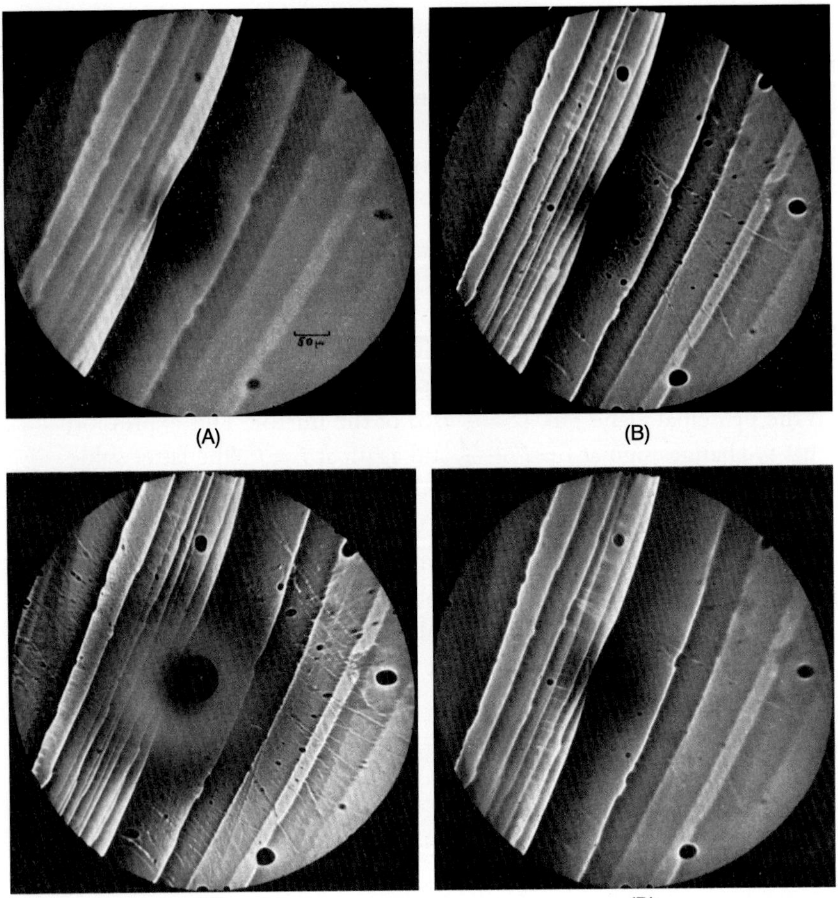

(A) (B) (C) (D)

Figure 45 Cleaved surface of rook salt crystal (Barnett & Nixon, 1967b), with changing specimen bias: (A) $V_b = +10$ V, (B) $V_b = -4.3$ V, (C) $V_b = -0.2$ V, (D) $V_b = +1.8$ V.

tern consisting of two interlocking comb structures; the regions between the (dark) copper bars are of high resistance. The periodicity of the structure on the left-hand side is 100 μm and on the right it is 200 μm. In Fig. 46A there is no voltage between the two combs and hence the contrast is entirely of topographical origin. In Fig. 46B alternate bars (i.e. those of the right-hand comb) are 0.75 V negative with respect to the remaining bars. The bars marked by arrows are disconnected and are used as references. It is seen that the effect of the electrical potential difference is to cause the more negative bars to expand at the expense of the more positive.

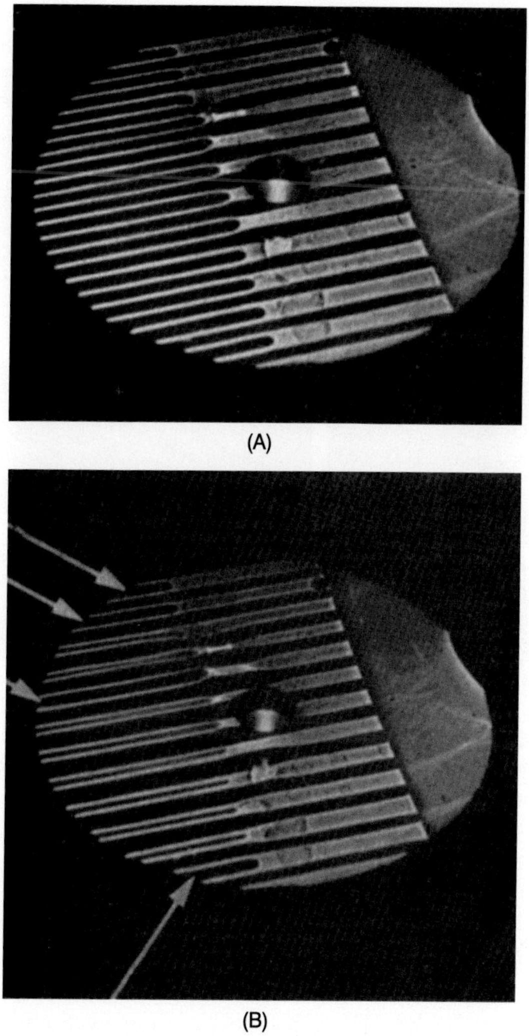

Figure 46 Interdigital electrode structure (Szentesi and Barnett). (A) Topographical contrast, (B) mixed contrast: right-hand comb 0.75 V negative with respect to left-hand comb.

Fig. 46A shows the characteristic image distortion which is a serious defect of shadow projection mirror microscopy. On the left-hand side of the micrograph, the bars appear to be much wider than the intervening spaces, even though in the actual physical structure, the bars (35 μm wide) are considerably narrower than the spaces. The theory of this unavoidable

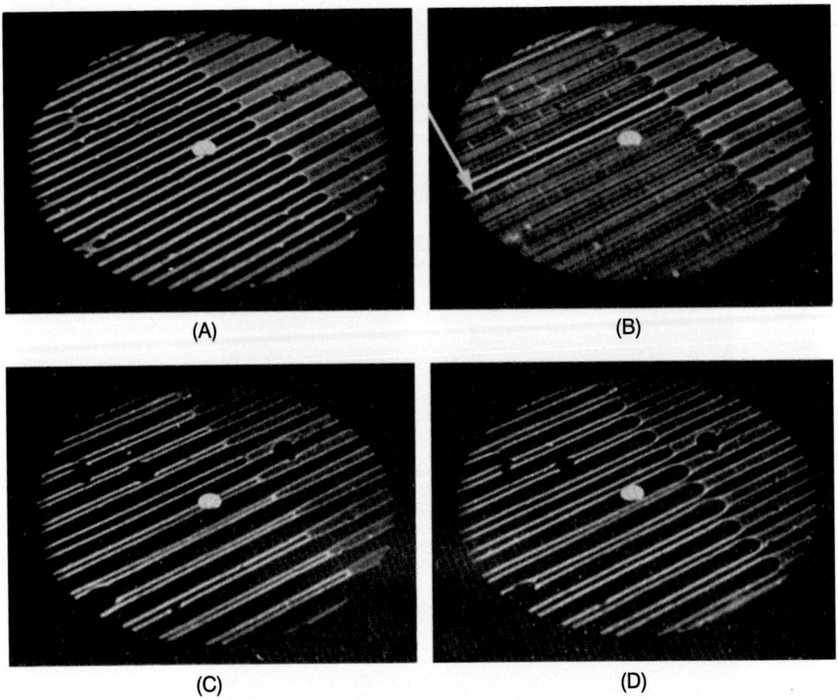

Figure 47 60 MHz stroboscopic images of an interdigital structure (Szentesi and Ash). (A) Topographical picture, (B) a.c. potential difference viewed from continuous beam, (C) and (D) a.c. potential variation viewed stroboscopically.

distortion, which exaggerates the area of topographical hills or regions of negative potential, has been worked out for periodic structures by Barnett and England (1968).

The problem of obtaining test specimens in which the contrast is purely of electrical origin can be largely overcome by using evaporated photoconductors on to which a suitable light pattern is focused. A system of this type has been realized at Imperial College by Barnett, Bates, and England (1969). If vitreous bismuth–selenium layers are used as specimens, the system operates as an infra-red image converter.

Alternating electrical potentials can be observed using the electron mirror, provided the beam is pulsed at a repetition frequency equal to the frequency of the surface potential variation. Fig. 47 shows an example of this stroboscopic technique (Szentesi, 1970); the specimen is an interdigital structure similar to that of Fig. 46. Fig. 47A is taken with a continuous beam, with no voltage between the combs. Fig. 47B shows a continuous

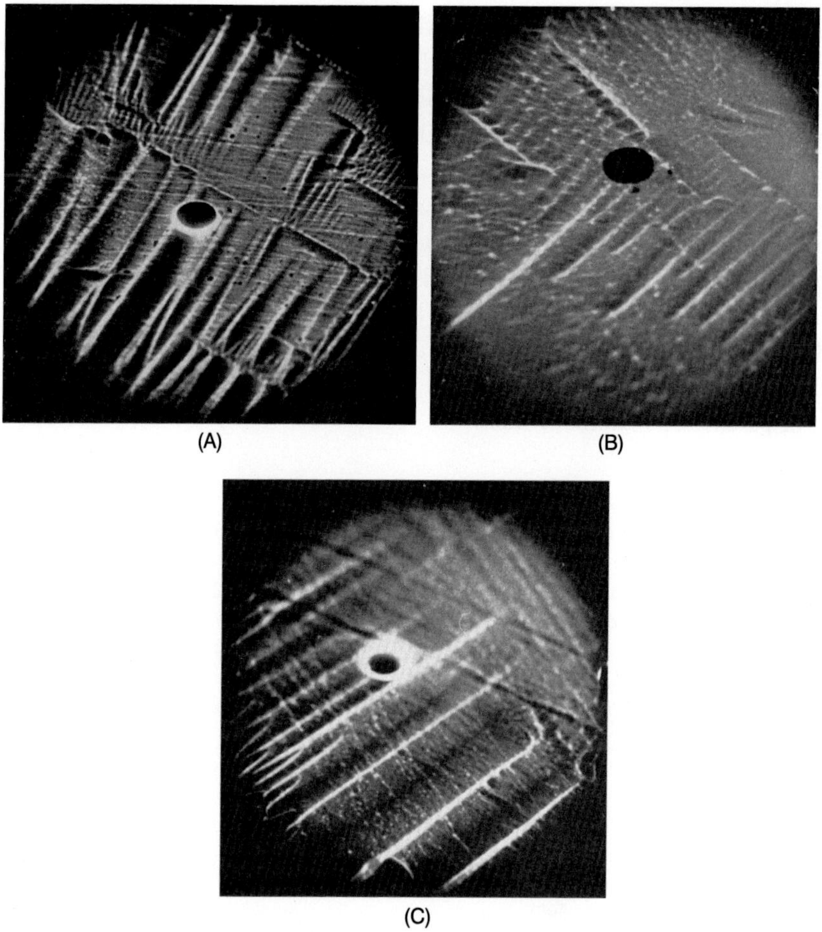

Figure 48 Prism face of cobalt crystal (Barnett & Nixon, 1964). (A) Dagger domains along a grain boundary, (B) a cluster of dagger domains (detail), (C) 180° domain walls.

beam image of a 66 MHz variation, in which the amplitude of the alternating potential difference between the combs is about 1 V. Figs. 47C and D are stroboscopic images of this variation. Fig. 47D shows the instant at which the left-hand comb is fully negative with respect to the right-hand comb, and Fig. 47C is half a cycle earlier. It is hoped to use this technique for the recording of ultrasonic holograms, using a piezoelectric crystal as the electron mirror.

Finally, some micrographs are shown which illustrate magnetic contrast in the mirror microscope. The specimen is the polished prism face of a cobalt crystal and the instrument an early three lens column (Barnett & Nixon, 1964). Magnetic contrast is largely due to the interaction between the radial velocity v_r of the electron and the normal component of the surface magnetic field. Eq. (63) implies that the most favorable conditions for magnetic contrast will be when f is as close to f' as is consistent with obtaining a sufficiently large field of view at the image. Furthermore, the contrast must disappear in a central region. This is seen in Fig. 48A where contrast due to topographical scratches is continuous through the region above and just to the right of the screen hole, while the magnetic features have faded out. Fig. 48A shows a row of dagger domains along a grain boundary; B shows details of a cluster of such dagger domains; C shows a system of parallel 180° domain walls.

REFERENCES

Abramowitz, M., & Stegun, I. A. (1965). *Handbook of mathematical functions* (p. 440). New York: Dover Publications.

Archard, G. D., & Mulvey, T. (1958). *Journal of Scientific Instruments, 35*, 279.

Barnett, M. E., Bates, C. W., & England, L. (1969). *Advances in Electronics and Electron Physics: Vol. 28* (p. 545). New York: Academic Press.

Barnett, M. E., & England, L. (1968). *Optik, 27*, 341.

Barnett, M. E., & Nixon, W. C. (1964). In *Proceedings of the 3rd European conference on electron microscopy: Vol. 1* (p. 37).

Barnett, M. E., & Nixon, W. C. (1967a). *Journal of Scientific Instruments, 44*, 893.

Barnett, M. E., & Nixon, W. C. (1967b). *Optik, 26*, 310.

Bartz, G., & Weissenberg, G. (1957). *Naturwissenschaften, 44*, 299.

Bartz, G., Weissenberg, G., & Wiskott, D. (1952). *Phys. Verh., 3*, 108.

Bartz, G., Weissenborg, G., & Wiskott, D. (1954). In *Proceedings of the 4th international conference on electron microscopy* (p. 395).

Bartz, G., Weissenberg, G., & Wiskott, D. (1956). *Radex-Rundschau, 163*.

Bethge, H., Hellgardt, J., & Heydenreich, J. (1960). *Experimentelle Technik der Physik, 8*, 49.

Bok, A. B. (1968). Thesis. Delft.

Bok, A. B., Kramer, J., & Le Poole, J. B. (1964). In *3rd European conference on electron microscopy*, A9.

Bostanjoglo, O., & Siegel, G. (1967). *Cryogenics*, 7, 157.

Forst, G., & Wende, B. (1964). *Zeitschrift für Angewandte Physik, 17*, 479.

Garrood, J. R., & Nixon, W. C. (1968). In *Proceedings of the 4th European conference on electron microscopy: Vol. 1* (p. 95).

Glaser, W. (1952). *Grundlagen der Elektronenoptik* (p. 321).

Haine, M. E. (1947). *Journal of Scientific Instruments, 24*, 61.

Henneberg, W., & Recknagel, A. (1935). *Zeitschrift für technische Physik, 16*, 621.

Heydenreich, J. (1966). In *Proceedings of the 6th international congress on electron microscopy: Vol. 1* (p. 233).

Heydenreich, J. (1970). In *Proceedings of the 7th international conference on electron microscopy: Vol. 2* (p. 31).

Hopp, H. (1960). Thesis. Berlin.
Hottenroth, G. (1937). *Annalen der Physik (Leipzig)*, *30*, 689.
Igras, E. (1961). *Bulletin de L'Académie Polonaise des Sciences. Série des Sciences Physiques*, *9*, 403.
Igras, E. (1962). In *Proceedings of the international conference on semiconductors* (p. 832).
Igras, E., Spivak, G. V., & Zheludev, I. S. (1959). *Soviet Physics. Crystallography*, *4*, 111.
Igras, E., & Warminski, T. (1965). *Physica Status Solidi*, *9*, 79.
Igras, E., & Warminski, T. (1966). *Physica Status Solidi*, *13*, 169.
Igras, E., & Warminski, T. (1967). *Physica Status Solidi*, *20*, K5.
Ivanov, R. D., & Abalmazova, M. G. (1968). *Soviet Physics. Technical Physics*, *12*, 982.
Knoll, M. Z. (1935). *Technical Physics*, *16*, 767.
Kranz, J., & Bialas, H. (1961). *Optik*, *18*, 178.
Le Poole, J. B. (1964a). In *Proceedings of the 3rd European conference on electron microscopy*, A6.
Le Poole, J. B. (1964b). In *Proceedings of the 3rd European conference on electron microscopy*, A8.
Le Poole, J. B. (1964c). In *Discussions on the conference on non-conventional electron microscopy*.
Le Rütte, W. S. (1952). Thesis. Delft.
Lenz, F., & Krimmel, E. (1963). *Zeitschrift für Physik*, *175*, 235.
Lukjanow, A. E., & Spivak, G. V. (1966). In *Proceedings of the 6th international congress on electron microscopy* (p. 611).
Maffit, K. N., & Deeter, C. R. (1966). In *Symposium "The mirror electron microscope for semiconductors"* (p. 9).
Mayer, L. (1955). *Journal of Applied Physics*, *26*, 1228.
Mayer, L. (1957a). *Journal of Applied Physics*, *28*, 259.
Mayer, L. (1957b). *Journal of Applied Physics*, *28*, 975.
Mayer, L. (1958). *Journal of Applied Physics*, *29*, 658.
Mayer, L. (1959a). *Journal of Applied Physics*, *30*, 1101.
Mayer, L. (1959b). *Journal of Applied Physics*, *30*, 252s.
Mayer, L. (1960). *Journal of Applied Physics*, *31*, 346.
Mayer, L., Rickett, R., & Stenemann, H. (1962). In *Proceedings of the 5th international congress on electron microscopy*, D-10.
Orthuber, R. (1948). *Zeitschrift für Angewandte Physik*, *1*, 79.
Recknagel, A. (1936). *Zeitschrift für Physik*, *104*, 381.
Ruska, E. Z. (1933). *Zeitschrift für Physik*, *83*, 492.
Schwartze, W. (1965). *Naturwissenschaften*, *52*, 448.
Schwartze, W. (1965–1966). *Optik*, *23*, 614.
Schwartze, W. (1966). *Experimentelle Technik der Physik*, *14*, 293.
Schwartze, W. (1967). *Optik*, *25*, 260.
Spivak, G. V. (1959). *Kristallografiya*, *4*, 123.
Spivak, G. V., Igras, E., Pryamkova, I. A., & Zheludev, I. S. (1959a). *Soviet Physics. Crystallography*, *4*, 115.
Spivak, G. V., Igras, E., Pryamkova, I. A., & Zheludev, I. S. (1959b). *Izvestiya Akademii Nauk USSR, Seriya Fizicheskaya*, *23*, 729.
Spivak, G. V., Ivanov, R. D., Pavluchenko, O. P., Sedov, N. M., & Shvets, V. F. (1963). *Izvestiya Akademii Nauk USSR, Seriya Fizicheskaya*, *27*, 1210.
Spivak, G. V., Prilegaeva, I. N., & Azovcev, V. K. (1955). *Doklady Akademii Nauk USSR*, *105*, 706, 965.
Spivak, G. V., Pryamkova, I. A., Fetisov, D. V., Kabanov, A. N., Lazareva, L. V., & Silinn, A. I. (1961). *Izvestiya Akademii Nauk USSR, Seriya Fizicheskaya*, *25*, 683.

Spivak, G. V., Saparin, G. V., & Pereversev, N. A. (1962). *Izvestiya Akademii Nauk USSR*, *26*, 13332.

Szentesi, O. I. (1970). Ph.D. thesis. London.

Szentesi, O. I., & Barnett, M. E. (1969). *Journal of Scientific Instruments*, *2*, 855.

von Borries, B., & Jansen, S. (1941). *Zeitschrift des Vereins Deutscher Ingenieure*, *85*, 207.

Wang, S. T., Challis, L. J., & Little, W. A. (1966). In *Proceedings of the 10th international conference on low temperature physics* (p. 150).

Wiskott, D. (1956). *Optik*, *13*, 481.

INDEX